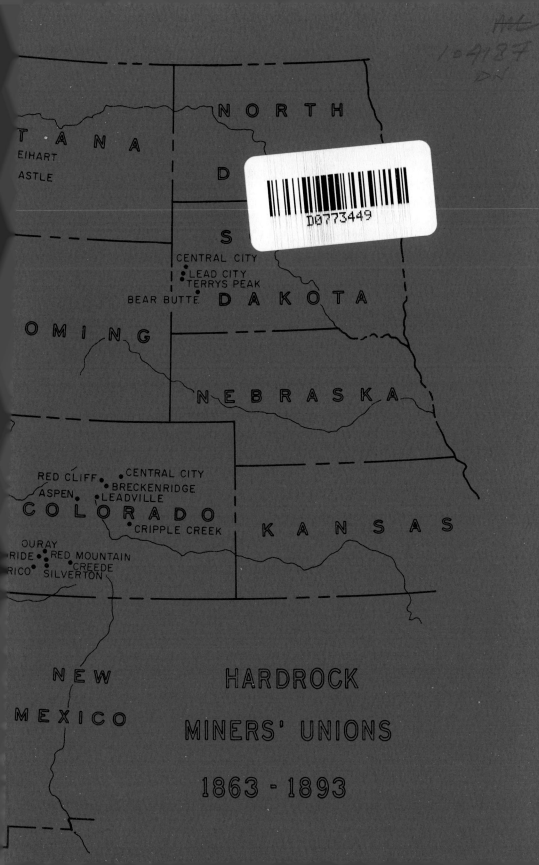

HARDROCK

MINERS' UNIONS

1863 - 1893

The Hardrock Miners

THE

HARDROCK MINERS

*A History of the Mining Labor
Movement in the American West
1863-1893*

Richard E. Lingenfelter

UNIVERSITY OF CALIFORNIA PRESS
Berkeley   Los Angeles   London
1974

UNIVERSITY OF CALIFORNIA PRESS
BERKELEY AND LOS ANGELES, CALIFORNIA
UNIVERSITY OF CALIFORNIA PRESS, LTD.
LONDON, ENGLAND
COPYRIGHT © 1974 BY THE REGENTS OF THE UNIVERSITY OF CALIFORNIA
ISBN: 0–520–02468–0
Library of Congress Catalog Card Number: 73–78547
PRINTED IN THE UNITED STATES OF AMERICA

Contents

Preface

What survives today of the story of the early western miners' unions may seem only a catalog of conflict and confrontation, but then as now that is what made the news and left the deepest scars on the memory. Indeed the importance of these conflicts cannot be denied, for only through them came the advancement of miners' rights. But, the more pervasive though less tangible feelings of strength and security, pride and benevolence, that the unions fostered among the hardrock miners find much less expression in the scraps of history and their importance seems diminished. These feelings are only suggested in the boastful demand that a silver miner is worth his hire in gold coin, in the melancholy memorial to a dead brother, or in the jubilant announcement of the annual union picnic. But in these feelings lies the spirit that is the very foundation of the labor movement and the ubiquitous force that gives it solidarity.

In the following pages I trace the history of the labor movement in the hardrock mines of the West through three decades, from the founding of the first miners' union on the Comstock Lode in 1863 to the federation of the western miners' unions in 1893. In so doing, I try to explore the relationship between the movement and the industrialization of the mines which sparked it, to follow the development of its organization, its activities and its goals, and perhaps most important, to glimpse a bit of the hardrock miner's life and aspirations, which the movement ultimately reflected.

Gathering together the scraps from which this history has been reconstructed would not have been possible without the generous aid of many individuals, libraries, historical societies, and state and union archives. I am particularly indebted to Edith Fuller and others in the Research Library, University of California, Los Angeles; John Barr Tompkins of the Bancroft Library, University of California, Berkeley; Robert D.

Armstrong of the Special Collections Library, University of Nevada, Reno; Harriett C. Meloy of the Library of the Historical Society of Montana, Helena; and Irving Dichter, secretary-treasurer of the International Union of Mine, Mill and Smelterworkers, Denver.

I am also very grateful to Ira B. Cross of the University of California, Berkeley; David F. Myrick of San Francisco; Stanton J. Peale of the University of California, Santa Barbara; Duane A. Smith of Fort Lewis College, Durango; and Robert W. Smith of Oregon State University, Corvallis.

Finally I would especially like to thank Rodman W. Paul of the California Institute of Technology, Pasadena, for his valuable comments and suggestions on the work as a whole.

The Hardrock Miners

The Hardrock Miner

The history of the early mining labor movement in the West is the history of the hardrock miner. It is the story of his hopes and trials, his joys and hardships, his aspirations and grievances, his successes and failures. It is shaped by his background, by the hazards of his daily work and by the changing technology and management in his trade. The goals of the movement are a tribute to his compassion and intellect; the successes are a monument to his cooperation and determination; the violence is a mark of his impatience and frustration.

Most of the men calling themselves miners in the early West, however, were not hardrock men working in the deep mines. They worked on the surface in the placers—panning, sluicing, and hydraulicking. They were not as highly skilled nor was their work as hazardous as that of the hardrock miner. They had come West in the great mining rushes to California in '49, Colorado in '59 and in many later rushes. They were driven from their farms, shops, and factories by hard times and drawn to the western gold fields by dreams of quick fortune, a fresh start, or one last chance. Most "saw the elephant" soon, quit mining, and settled down to a farm or shop in the West. Those who lingered in the placer mines still nurtured the dream that had drawn them to the West. They were predominantly Americans and Chinese. Until the late 1870s, in fact, most of the Chinese in America worked in the placer mines, frequently buying up abandoned claims and reopening them at a profit.[1]

The opening of deep, lode mines required much more capital and skill than placer mining, and consequently deep mining developed much more slowly. An extreme view of the difference between placer and deep mining was drawn by the editor of the San Diego *Union*. "We are glad to see placer mines developed, as these are the mines of the people. They are the mining regions where the individual is thrown on his own resources and personally attains his successes. With quartz and deep mines the case is very different. The great attendant expenses of such mines convert the individual into a factor in the corporation and degrades the sturdy miner into a drudge in the drift, toiling at so much per day, while his bosses—his owners in fact—reap the great profit of his endeavor." The history of the western mining labor movement is the story of the hardrock miners' struggle against the fulfillment of that dismal fate.[2]

Deep mining in the West began in the gold-bearing quartz of the Mother Lode and Grass Valley in the 1850s, grew in the rich silver lode of the Comstock in the '60s and '70s, and matured in the lower grade, base metal ores of Eureka, Leadville, Butte, and the Coeur d'Alene in the '80s and '90s. The yearly bullion product of the western deep mines rose from a few million dollars in the beginning to tens of millions with the opening of the Comstock and reached more than a hundred million by the end of the century. During this time the number of hardrock miners rose from hundreds in the '50s, when they made up less than a hundredth of the western miners, to several thousand in the early '70s, when they were about a tenth of all the miners, to over thirty thousand in the early '90s, when they finally outnumbered the placer miners. With the growth of deep mining came ever increasing absentee ownership and industrialization and an ever increasing need for collective bargaining and strong miners' unions.[3]

Among the hardrock miners the native-born American and the Chinese were in the minority. When he did engage in deep mining the American rarely did so for very long at a time. Still saddled with the dream of striking it rich, he was generally a restless sort, who traveled in that nomadic band always the first to arrive in a new camp and the first to leave. He was not content to work for others any longer than it took to raise a stake to prospect on his own hook. With a cayuse or mustang, a pick and shovel, and a sack of selfrising flour, he was off to seek his fortune in every distant hill. Even when he made a strike and located a claim of his own, he was likely only to open it enough to strip out a little highgrade and make it attractive to some capitalist. It was said that the typical American miner would tear up the streets of Pandemonium if he could get ten cents to the pan, and the only chance he had of getting to heaven was to start a gold rush there.[4]

The Chinese was barred from working as a hardrock miner in the larger camps by his prejudiced brothers, but he did get work in some of the more isolated deep mines. Coming primarily from the delta farming communities around Canton, he, like the American, had no previous experience in deep mining. But in tunnel work on the Central Pacific Railroad, in the Mariposa and other deep mines along the Mother Lode, and in the quicksilver mines of California, he demonstrated his ability to learn the skills of the trade. His white employers found to their surprise that he could "surpass white men employed in the same mines." But they still paid him much less and were annoyed that he had so little faith in their promises that he would stop work if not promptly paid. His employers also complained of the frequency of his religious festivals and holidays, which failed to coincide with those of Christian America. His white co-workers, however, raised the noisiest complaints about his employment and made opposition to the Chinese an important part of the mining labor movement.[5]

The most experienced miner in any deep mine in the West before the early 1860s was likely to have been the Mexican, who had worked in the gold, silver, or copper mines of northern Mexico. He may have come with the first rush to the lands lately taken from Mexico, and, after a brief spell in the placers, turned again to lode mining. There his practical experience in mining and milling enabled him to work a claim of his own to good profit or to get work at a good wage. With the discovery of silver on the Comstock skilled Mexican miners were in heavy demand. Even a vaquero from the cow counties of California could draw good money on the Comstock in the early years when any man who spoke Spanish was presumed to have mined all his days in the richest silver mines of Mexico. But these good times were short-lived. Soon even the skilled Mexican miner found his methods denounced as primitive and his place filled by the tin miner from England. By 1880 there was not one Mexican among the two thousand miners employed on the Com-

stock Lode. Although the Mexican miner could still find a job
in the mines of New Mexico, Arizona, and parts of California,
even there he had to settle for discriminatory wages of only
50¢ to $1.50 a day, whereas his less experienced gringo brothers
drew from $2 to $3.[6]

Starting in the mid 1860s Cornish and Irish made up the majority of the hardrock miners in the larger camps. The Cornishman, or "Cousin Jack," was raised in the trade and was recognized as "the most skillful foreign miner that comes to our
shores." He was, in fact, the miner's miner. Hard times had
driven him from the tin, lead, and copper mines of Cornwall,
but he came to the western mines with few illusions. Higher
wages, not dreams of quick fortunes, attracted him and he came
to work. His greater skill gave him an obvious advantage in the
competition for jobs—an advantage that occasionally sparked
resentment among his American and Irish fellow workers. His
clannishness and the bitter legacy of English-Irish conflict further heightened the antagonism. This animosity between the
skilled Cornish miners and the initially unskilled Irish and
American miners was, in fact, the most explosive internal threat
to the mining labor movement in the West. It surfaced in divisive
ways throughout the West: on the Comstock at the very start
of the movement when Cornish were hired effectively as strikebreakers; in Gilpin County, Colorado, where Irish and American miners refused to join a predominantly Cornish union; at
Cherry Creek, Nevada, where a union was organized "to counter-
act the manifest preference given by mine managers to Cornishmen," and at Butte where Cornish-Irish rivalry for control of
the union reached proportions so grotesque that it ultimately
split the union. Despite such extreme aberrations, however, the
Cousin Jacks for the most part got along quite well with their
fellow miners, and in marked contrast with a solid antiunion
tradition in Cornwall they became the leaders of the mining
labor movement in the West.[7]

Hard times also brought the Irishman to America as a laborer
and his search for work drew him to the western mines where
the wages were good. There, like the American and Chinese, he
graduated from laborer to miner, quickly learning to handle the

pick and shovel with "much ease and grace." He was often a devil-may-care sort of fellow, who would divide his last crumb and give his last dollar to a friend in need. But when he got into a row, a pick handle became "a terrible weapon in his hands." He also generally took a deeper interest in politics than most of the other foreign-born miners.[8]

In the Comstock mines the Cornish and Irish each made up a third of the hardrock miners; only a fifth were native-born Americans; less than a tenth were Canadians; and the remainder included a scattering of Scots, Germans, Welsh, French, Swedes, Italians, Austrians, Prussians, Portuguese, Swiss, and Danes. The Comstock miners were fairly representative of hardrock miners in most of the silver camps in the Great Basin during the latter half of the last century. In the bordering gold camps, however, the makeup of the miners sometimes differed significantly. In Arizona, New Mexico, and parts of California Mexicans made up a major fraction of the crews and in the Rocky Mountains southern and eastern Europeans were more numerous. Many of the mining superintendents purposely hired men from as wide a range of nationalities as possible in the hope that cultural and linguistic differences would make union organization more difficult. But despite national rivalries the hardrock miners generally worked side by side with remarkable harmony —so long as Chinese, Mexicans, and Blacks were excluded. The common bond of the miners' union did much to keep the peace among them.[9]

In physical build the hardrock miner was not overly brawny, for skill rather than strength was the mark of his trade. In 1880 the average miner on the Comstock stood five feet nine inches and weighed a hundred sixty-five pounds. His average age was thirty-six. Inwardly he was calmly proud of his occupation, but outwardly he was a modest man. His ability to support himself, if necessary, on a small high-grade claim of his own gave him a measure of independence. This option was important in supporting higher wages in the western mines, for as long as the miner thought he could do better on his own he rarely would work for less.[10]

The hazards of his work generally made him thoughtful and

dependable, for the safety and even the lives of his fellow workmen often hung on the care with which he did his work, and
his own life hung in turn upon their work. A single careless
prop, a defective bolt or timber, a carelessly swung hammer,
an improperly loaded charge, any neglect or lack of thorough-
ness, any laziness or ignorance was almost certain to bring injury and possibly even death.

At the same time the hazards of his work also made him
something of a gambler, in practice as well as spirit, for every
day he gambled with his life in the uncertain depths of the mine.
By comparison with such a gamble, he had little to lose at the
gaming tables and he frequently turned to them for amuse-
ment, betting with abandon. He gambled as freely with his
future. His high wages gave him a good opportunity to save a
fair fraction of his earnings. But he was more likely to stake himself to a prospecting trip for the chance of making a rich strike,
or to gamble in mining stock and lose. With few exceptions the
Comstock miners staked their savings on the rise and fall of
mining stocks. None knew better than they the emptiness of
stock "deals" and the uncertainty of actual "strikes," and none
were more aware of the power of the manipulator and the weakness of the investor. None saw the odds against them more
vividly and yet none accepted these odds more readily. For
their employment in the mines with its apparent opportunity
of learning conditions firsthand gave them a delusive advantage.
Even when they knew the true condition of a mine and knew
that its stock was being manipulated, they would still try to
share in the profits of the deal and they would give their mite
to swell the bubble. Thus few Comstock miners saved much
money for the stock exchange swallowed up all their surplus.
Indeed one prominent mine owner and stock operator remarked
that he didn't care how high miners' wages were, for whatever he paid them over $3 a day would come back to him
through the stock market.[11]

In the popular literature of the time the western miner was
pictured in a sombrero, a red shirt draped with cartridge belts,
an obtrusive army revolver, and trousers tucked into his books.
In fact, in the mine he worked stripped to the waist and above

ground he was a conservative dresser, preferring plain, sober-colored suits of durable material though "not faultless in fit." Nonetheless his street clothes were better than the holiday dress of the average American laborer.[12]

The hardrock miner in the larger camps was often as not a married man. Many a miner in the smaller more ephemeral camps may also have been married but his wife most likely had not come West with him. For if she did she was constantly faced with the precariousness of the miner's life and the tragedy that might befall her family. "It requires the highest moral courage to be a 'miner's wife,'" one westerner noted, for every farewell kiss reminded her that he might return a cripple or a corpse, torn and bleeding from some horrible accident, mangled by a blast, crushed by a cave, or mutilated beyond recognition by a fall down a shaft. "Such contemplations are not calculated to make the life of the 'miner's wife' a happy one. . . . Yet she is cheerful and exerts effort to make home pleasant and comfortable, and to banish from her mind the terrible dread of what is almost certain to happen."[13]

If his wife accompanied him, the miner usually owned his own home, which cost him anywhere from a few hundred dollars to over a thousand, representing from a few months' to a year's wages. His home was probably a plain board cabin, neither whitewashed nor painted—even less pretentious than his dress. In outward appearance it was decidedly inferior to most workers' houses in the East. But inside it was usually well furnished and when there was a woman in the house the rooms were "transformed into cosy lodgings by bright-colored curtains, soft carpets, prettily-figured wall-paper, and the hundred lesser additions which make even a rude cabin interior pleasing to the eye." Here the miner and his wife could find pleasure and comfort in the evenings, and "gather their little ones around them and listen to their innocent prattle, forget the cares of life."

But the married miner also had the financial burden of his family and other disadvantages. Some mining superintendents felt that men with families were "less vigorous, less energetic and less daring"—that is, less willing to risk their lives for the company. Thus in lean times the married men were frequently the first to be dropped from the roll. This worked doubly hard on them, since they could least afford to move on to other camps in search of work. Still, there were exceptions to this discrimination, as some more compassionate managers kept less able men solely out of consideration for their families. Also the single men often left a district voluntarily in order to give a better chance of employment to men with families, and working members of the union allowed unemployed members to take a share of their shifts to help tide them over.[14]

For most of the miners, however, their only home was a cheerless room in a boardinghouse, where high prices were the rule and homey accommodations the exception. There they stayed, dependent on the sufferance of a landlady, who looked upon her customers as created solely to occupy the rooms and pay their rent. The one who was the least trouble stood highest in her favor. Certainly there was nothing attractive about the best of boardinghouses to entice a man to spend his leisure hours in them. A single bedstead with a washstand and a common chair was the usual complement of furniture, and the only comfort to be extracted was a hasty plunge under the blankets, seeking the oblivion of sleep.[15]

But the dreariness of his room was the least of the miner's complaints about his boardinghouses and bunkhouses. Much less tolerable was the sloppy grub on flyspecked plates, the filthy, varmint-infested beds and bedding, and the rooms—cold, wet, and drafty in the winter; hot, smelly, and fly choked in the summer; and noisy year round. The miner paid over a dollar a day for such board and lodging, and his only recourse was to grumble. One Utah miner set down in song his gripes,

> ... about the boardinghouse up at the Apex Mine,
> Where they make us Zion biscuits just as hard as any slug;
> You would of died, had you of tried old Curly's awful grub.
>
> The coffee has the dropsy; the tea it has the grippe.
> The butter was consumptive, and the slapjacks they had fits;
> The beef was strong as jubilant; it walked upon the floor.
> The spuds got on their dignity and rolled right out the door.
>
> The pudding had the jimjams; the pies was in disguise.
> The beans came to the table with five hundred thousand flies.

The hash was simply murdered, just as hard as 'dobe mud.
We howl, we wail, our muscles fail on Baxter's awful grub.[16]

Not surprisingly the single miner spent his idle hours in the street, the saloon, the hurdyhouse or the whorehouse. An occasional strolling troupe of players or a holiday ball were momentous events that brought brief relief from the monotony of camp life. But their appearance was rare and they were soon gone, leaving the miner little resource but the saloons, whose bright lights, cheerful fires, cosy rooms, and attractive games drew in the crowds from the street as soon as night fell. There the miner could settle into a comfortable armchair, prop his feet on the stove, and while away the hours with a choice circle of friends, discussing national affairs or the merits of the last barroom encounter, while someone treated with an occasional round from the bar. If this was too tame, there was always the excitement of the card table with a game of pedro or seven-up for a trifling stake and an incidental round of drinks, or of the faro table, even as a spectator, if some reckless player was setting them in, trying to break the bank, or in the keno room, where he could buy a card and a few minutes of hope for a quarter. For variety he might wander across the street, where the music floated out on the air to entice him to the allurements of the hurdyhouse. Here he could catch a comely partner in a firm embrace and whirl away in a voluptuous waltz, thread the mazes of a cotillion, or relieve his pent-up feelings in an impromptu breakdown, before the orchestra ended the dance in a grand crash and he paid the price of his indulgence at the bar. But as the hours slipped away, he ultimately had to face that lonely boardinghouse room once more or buy a few hours in a crib.[17]

The miners' consumption of beer, wine, and hard liquor was considered "miraculous" by one eastern visitor. But so too was their capacity for holding it and he concluded with awe that "custom has made it a property of easiness to take glass after glass without visible effect." Surprising, too, to the easterner, the crime rate in the larger camps was lower than in most eastern cities. Only in the more remote camps did the itinerant gunmen congregate to have their "man for breakfast."[18]

Despite the miner's indulgences and vices, he was generally

a responsible citizen who took an active part in his community. He delighted in political debate, excercised his vote at nearly every opportunity, and elected his fellow miners to such local offices as alderman, justice of the peace, and county sheriff, and to seats in the Territorial and State Legislatures and even the United States Congress. He also gave financial support to the schools, the church, charities, and his union.

But the hardrock miner was not measured by these things; he was measured at his work in the depths of the mine by his physical skill and his technical knowledge, his cool-headedness and his courage. For in the "dreary depths of the darksome mine" a man was assayed far more critically and tempered more lastingly than he was beside the saloon stove, at the faro table, on the hurdyhouse floor, or in the crib. To understand the hardrock miner then, is to understand the depths of the mine.[19]

There was usually a simple order to the seeming labyrinth of underground workings. Since the lode or vein often dipped rather steeply into the earth, a shaft was sunk to open it at depth. On a steep hillside an adit, or tunnel, served the same purpose. Every hundred feet of depth generally marked a level where a drift or gallery was run paralleling the lode. These were from four to six feet wide and seven to eight feet high. If the rock was "slabby" or loose, timbering was required to support the roof and walls against caving. Crosscuts, also about a hundred feet apart, ran at right angles from the drift to cut the lode. If the crosscut struck paying ore in the lode, either cross drifts or a stope were opened in the lode to bring down the ore. The ore and waste rock were removed in hand or mule cars running on tracks laid through the crosscuts and drifts to the shaft. The rock was then either hoisted to the surface or run out the tunnel. Since moving the rock was costly, waste rock was dumped in abandoned portions of the mines if possible. The labyrinth was further complicated by winzes and inclines connecting various levels. In addition to the main shaft, smaller air shafts might be sunk for ventilation. If, however, more than one company was working on a lode, connections between drifts of adjacent mines provided a draft of air through the main shafts. The direction of the draft in a mine usually remained constant for years; only a fire or a new shaft or connection could change it. The natural draft was rarely adequate for ventilation in deep mines, however, and blowers were needed to force fresh air through a network of pipes to the deep workings of the mines.[20]

To the visitor in the mines there was often "something weird, something fascinating about the deep, solemn hush, the death-like stillness, and the grim tomb-like quiet of the long, lonesome drifts and galleries and the deep hollow reverberating chambers far down beneath the ground." Alone in some unfrequented portion of the mine with only a flickering candle for a companion even the practical miner sometimes felt a tinge of superstition.[21]

But more often the physical hardships of work in the deep mines forced any romantic or superstitious notions out of the miner's mind. Eliot Lord wrote forcibly of the oppressive conditions in the hot levels of the Comstock mines, where the heat alone could kill a man.

View their work! Descending from the surface in the shaft-cages, they enter narrow galleries where the air is scarce respirable. By the dim light of their lanterns a dingy rock surface, braced by rotting props, is visible. The stenches of decaying vegetable matter, hot foul water, and human excretions intensify the effects of the heat. The men throw off their clothes at once. Only a light breech-cloth covers their hips, and thick-soled shoes protect their feet from the scorching rocks and steaming rills of water that trickle over the floor of the levels. Except for these coverings they toil naked, with heavy drops of sweat starting from every pore. . . . Yet, though naked, they can only work at some stopes for a few moments at a time, dipping their heads repeatedly under water-showers from conduit pipes, and frequently filling their lungs with fresh air at the open ends of the blower-tubes. Then they are forced to go back to stations where the ventilation is better and gain strength for the renewal of their labor.[22]

Tons of ice were sent down daily into the mines where the half-fainting men chewed fragments greedily to cool their parched throats, and carried lumps in their clenched hands through the drifts. Ice and water were consumed in extraordinary quantities. Three gallons of water and ninety-five pounds of ice was the average daily consumption of miners employed in the hottest workings of the Comstock. If power drills had not been used the work of exploration would probably have ceased. To penetrate hard rock while breathing such an atmosphere would have taxed human endurance too greatly; even to cut out the decomposed feldspar with light steel picks was a painful labor. At some stopes in the bonanza ore of the California and Consolidated Virginia mines four miners could scarcely do the ordinary work of one man in a moderately cool drift.

In some of the mines the men had to contend not just with the heat, but with scalding water as well. An incline in the Savage mine tapped a hot stream that chocked the air with steam. Men could stand for only a few minutes at a time near this hot fountain, breathing the suffocating vapor till they staggered back half blinded and bent double in agonizing cramps. When the pain was so great that they began to rave and jabber incoherently, their companions would quickly carry them to the coolest place on the level and rub them vigorously until their checked perspiration began to flow again and they regained their senses. Sometimes, however, the effects were more lasting. After such an ordeal one man remained in a dazed condition for several days, babbling like an infant, and never fully regained his memory.[23]

The heat simply killed others outright. Even repeated warnings were not always heeded, for often men did not realize their danger until their lives were sacrificed. So when a miner, working in the same rock furnace where another had died only a month before, turned a deaf ear to his companions who urged him to go to a cooling station, they were not surprised to see him drop his pick and fall dead. To lie gasping with swollen veins and purple face on a hot rock floor till the dull eyes are glazed in death was a dreadful fate, but the places of the dead were filled before their bodies were buried.[24]

The heat in the deep levels of the Comstock was extreme but it was not unique. The hardrock miner labored in temperatures well above 100° F. in the deep levels of many mines throughout the West. And heat took other tolls on the miner's health in addition to heat exhaustion. His rapid emergence, half-clothed and sweating, from the hot depths of the mine into the chill mountain air too often sent him home half choked by pneumonia and spitting blood. Agitation on the Comstock and elsewhere finally led to the introduction of changing or drying rooms, where the miner could shower and put on dry clothes before braving the elements. In mines with free milling gold, the owners also found such rooms an effective curb against highgrading.[25]

The heat and ice water also wracked many a miner's stomach; made his normal diet unpalatable; and sent him craving such delicacies as fruits and highly seasoned dishes, pickles, salads, pig's feet, hams—almost any food with an acid or salt flavor. On the Comstock, the miners demanded and got the best supplies in the market to satisfy their squeamish taste.[26]

But more common than heat, the hardrock miner suffered from foul air, poisonous gases, and dust in every poorly ventilated mine. The western mines were, in fact, notorious for their lack of adequate ventilation. The managers simply went down after the ore and the mines were usually left "to ventilate themselves." Even Rossiter Raymond criticized his fellow mining engineers for showing "a degree of carelessness" in ventilation that "no one would dare to be guilty of in a colliery." But he also noted that too often the miners were willing to accept such conditions with only a grunt that " 'the air is bad' in this or that level, very much as one would speak in helpless resignation about the weather out of doors."[27]

The air in the deep mines was fouled both by the stench of excrement and by the accumulation of carbon dioxide. In deference to his own nose the miner usually defecated in some unworked part of the level and the stench was more likely unpleasant than unhealthful. Carbon dioxide, however, could cause more serious problems. His every breath added to its accumulation in the mines, as did his candles, the slowly rotting timbers, and the carbonated ground waters. If too much carbon dioxide accumulated, the miner would lose consciousness and, unless carried to safety, he would suffocate within a few minutes. Even lesser concentrations could cause headaches and feelings of weakness, drowsiness, and dizziness that made him far more accident prone. A sizable fraction of mine accidents doubtless resulted indirectly from excessive carbon dioxide. A similar but more acute danger existed from carbon monoxide generated by burning candles and explosives.[28]

Breathing the residual vapor of unburned nitroglycerine after an explosion also caused painful, throbbing headaches, nausea, and vomiting. Because of these effects the miners of Grass Valley fought a prolonged battle with the mine owners to outlaw nitroglycerine in the mines. It was found, however, that after a short time a miner could develop a tolerance to the effects but even then prolonged exposure could lead to blindness and in severe cases insanity. The only real solution was adequate ventilation.[29]

The miner working in silver-lead ores suffered additionally from lead poisoning, or "leading," as a result of inhaling lead compounds in the dust. This was in fact the main cause of illness among miners in the silver-lead mines of Utah, Leadville, Wood River, and the Coeur d'Alene. The leaded miner was seized by severe gastric and abdominal pains which might persist as long as a day, sometime accompanied by a shakiness in the hands and fingers. He usually had to spend a week or two in the hospital on large doses of magnesium chloride before he could return to work.[30]

The most general affliction of the hardrock miner, however, was silicosis, caused by breathing the fine silica or quartz dust from drilling and blasting. Throughout the last century, half or more of the miners in the dry levels of western mines were afflicted with silicosis. Silica dust in the lungs causes fibrous nodules to develop which impair breathing and can ultimately be disabling. In advanced stages silicosis is frequently complicated by tuberculosis. In the nineteenth century the disease was known as miner's consumption, miner's asthma, or simply the miner's disease, and its cause was not understood. It was variously blamed on such factors as the dark, damp atmosphere of the mines or the poisonous gases from nitroglycerine. As late as 1911, physicians in Butte, Montana, where about half the miners were still suffering from it, testified to a legislative committee that dust was "absolutely not responsible for the miner's disease." As a result many a miner "lingered down to the tomb in the never relaxed clutch of miner's consumption" before the dust was finally controlled by wet drilling and effective ventilation.[31]

But the physical endurance necessary to work under such conditions was only one of the necessary qualifications of the hardrock miner. So too was the ability to swing an eight-pound hammer or manage a machine drill accurately in cramped quarters and poor light. The hardrock miner had to judge how many shots would be required to bring down the rock, where to put each hole, and how deep to drill them, so that they would be most effective. He had to have enough knowledge of mineralogy to recognize often deceptive ores when struck, and to separate them when stoping and hand sorting. He also had to gauge the strength of the rock to determine the timbering needed and he had to know enough rough surveying to keep the workings lined up. In larger mines some of these decisions were made by specialists, but in most mines all were a part of the hardrock miner's job.[32]

The miner usually worked a shift of from eight to ten hours. In mines working around the clock the shifts began at seven in the morning, at three in the afternoon, and at eleven at night. The miner ate a midshift meal in the mines wherever he happened to be working. Beginning his day, he reported to the shift boss and was lowered down the shaft. In a large mine he might descend at 400 feet a minute, crowded with a dozen other men of a 5-by-5-foot platform or cage; in a smaller mine he might have to stand on the rim of an iron bucket holding on to the hoisting rope. When he reached the level where he was to work, he picked up his candles, drills, hammer, pick, or shovel at the station and went to the spot assigned him. The candles were either attached to his cap or mounted in steel holders that could be driven into a crack in the rock. In some mines the miner was rationed to three candles per ten-hour shift and if he used them too fast he had to grope through the last of his shift in darkness.[33]

Whether the miner was sinking a shaft, running a drift, or bringing down ore at the breast, his drill work was essentially the same. If he worked alone, single-hand drilling or "single jacking," he put in a hole twisting a ¾ inch drill in one hand

and swinging a three-pound hammer in the other. Cousin Jacks
had a deep-seated dislike of single jacking and their oppostion
to it was a contributing factor in the labor dispute at Grass
Valley when it was used there. More likely the miner worked
with a partner, double hand drilling or "double jacking." He
wielded an eight-pound hammer with both hands while his
partner twisted a 1-inch drill. With each blow he hit the drill
he made a hissing noise, both to warn his "pard" that the hammer was coming and as a vent to his concentrated energies.
With equally good time and precision—for if he did not a broken
forearm would be the consequence—his "pard" brought out the
drill about an inch and turned it a little. It required strong
nerves to stand there and let a big hammer whiz by your ear,
remembering at the same time to keep time in "twisting" the
drill. The danger was heightened by the fact that the drill head
was a hard target to hit in the faint light of a couple tallow
candles. Periodically they stopped to clean out the hole and
change places.[34]

Machine drills, powered by compressed air, were introduced
in the Comstock mines in the early 1870s, shortly after their
invention. Without them deep mining in the hot levels would
have been impossible. Still they did not come into general use
throughout the West until the present century. The most popular early machines were the Burleigh and Ingersoll drills. The
drill bit was attached to a reciprocating piston driven by compressed air. A single air compressor located on the surface usually provided the air piped to all of the drills in the mine. The
compressed air also served to ventilate the working areas and to
power small hoists, pumps, and blowers. The machine drills
were, however, very heavy and cumbersome, and they required
two men to operate them, even though they were supported by
a tripod or a vertical bar set between jackscrews.[35]

When the hole was drilled to a sufficient depth, perhaps four
to five feet, it was cleaned and loaded with explosive. Black
powder was used before the introduction of Alfred Nobel's
dynamite in the late 1860s. Dynamite sticks were a foot long
and contained about 40 percent nitroglycerin held in an ab-
sorbent. The miner carefully cut open each stick with a knife

and gently forced it into the hole with a piece of wood. He added a fulminate of mercury blasting cap and a fuse, and then tamped clay into the remainder of the hole. After checking the nearby braces and removing his tools, he called out, "Fire one hole on the breast," or wherever it might be, to warn others nearby. He then lit the fuse and retreated a safe distance. In smaller mines holes were fired at any time but in larger mines they were all generally fired between shifts or during meal breaks to allow more time for the smoke and dust to clear. A miner using hand drills could bring down about a ton of rock a day but this varied considerably with the hardness of the rock.[36]

When the smoke and dust had cleared, the shoveler, or mucker, came in to remove the rock brought down by the blasts. But first he carefully checked the rock overhead and barred down any loose pieces to make the area safe to work in. The mucker did not need the same knowledge or skill as the miner, but he was often effectively an apprentice, learning the trade and working up to being a miner. His job was to load the broken rock into cars and hand sort the ore from the waste rock. One man could muck for more than a dozen miners, as he was expected to load about sixteen one-ton ore cars during a shift. To meet this quota he frequently developed a knack for wedging large blocks in the bottom of the car so as to make as much empty space as possible, but he had to be careful not to get caught. When a rock was too large for the mucker to handle, a miner known as a blockholer was called in to put a hole in it and blast it.[37]

Most of the miners and muckers in the western mines worked for wages of $3 to $5 a day. But some were hired on contract or tribute. Contract work was bid for at so much per running foot of tunnel, drift or shaft. The contract was usually for labor only and the company provided the candles, tools, powder, fuse, and timber. The earnings of the miner depended on his skill in judging the hardness of the rock and if it turned out to be harder than he expected he ended up with a meager wage indeed. If wages were high in a district few miners risked contract work. Occasionally, however, a resourceful if unscrupulous miner could make good pay even from a bad contract, if he was not caught. Dan De Quille told of such an instance involving some roving miners who took a contract at a modest rate to extend a tunnel 10 feet, only to discover that the rock was so hard the contract would not even pay their grub. Faced with the dilemma, they soon concocted a scheme. When they took the contract, the owner had set a wooden plug in the wall of the tunnel to mark a point from which to measure the new work without having to survey the whole length of the tunnel. Thus the contractors simply pried out the plug, set a charge in the hole to obliterate all trace of it, and, measuring back a generous 10 feet, reset the plug in the wall. After loafing around out of sight for an appropriate length of time, they went to the owner and collected their money, complaining mightily about the hardness of the rock.[38]

Dishonest superintendents also worked the contract system with equal facility. They let fat contracts to their friends without competitive bidding for two or three times the cost of the work. The friends then sublet the contracts to miners on competitive bids and split the profits with the superintendent. When the manager of the British-owned Richmond Mining Company let every contract to a small circle of friends, the Ruby Hill Miners' Union exposed the racket and struck for competitive bidding.[39]

Tributing was the most popular system among the Cornish miners in England, but in the western mines it was employed very selectively and generally to the disadvantage of the miner. A tributer was allotted a certain area in the mine, called a pitch, which he could work. For each ton of ore he brought down, he received a percentage of the assay in excess of some minimum value. The company, however, never let a tribute on a known ore body assaying much more than the minimum, and frequently used tributing only to prospect unpromising areas of the mine. The rates varied with the nature of the ore. Tributes in the Eureka, Nevada, mines in the late 1870s paid 10 percent to 15 percent of the assay above $40 a ton. Thus if one man could break down about a ton of ore a day, it had to assay better than $70 a ton before he made $4 wages. Most of the miners on tribute

averaged little better than half that earned by those on wages
and they generally took a tribute only when wage work was
scarce. The gamble of striking paying ore, however, always
lured some into tributing. But in nearly a decade of tributing at
Eureka, only two tributers struck bonanza ore and when they
did the companies promptly dissolved their contracts and took
the ground away from them before they could turn a profit. The
tributers sued but lost. The miners' unions frequently opposed
the system and in several camps the companies successfully
broke the unions by hiring men exclusively on tribute.[40]

In addition to the miners and the muckers there were carmen
to run out the ore, timbermen to brace up the slabby ground,
engineers and mechanics to operate and maintain the boilers,
hoists, air compressors, pumps, and other machinery, black-
smiths to reforge the dull drills and picks, and messengers to
carry tools, orders, water, and ice to the miners.

The men on each level were supervised by a foreman. He reported to the shift boss, who in turn reported to the superintendent or manager. The foreman was responsible for getting
as much rock broken as possible, and more than once the miners
struck to demand the firing of too zealous a foreman. One mining camp Moses laid down the following Commandments to
his crew.

I

Thou shalt not slumber late in the morning, but shalt rise ere it
is day and break thy fast, for he that goeth to the mine late getteth
no candles, causing the transgressor to grope in darkness and the
shift-boss to indulge in profanity.

II

Thou shalt not take up thy position in the center of the cage when
descending or ascending the shaft, neither shalt thou appropriate in
thy person more room than the law allows, for thou are but of little
consequence among a whole cage-load of men, no matter what thou
thinkest to the contrary.

III

Thou shalt not hesitate on the station, or smoke thy pipe and talk
politics with the pumpman, for verily the shift-boss might suddenly
appear, and heaven help thee if he findeth the chutes empty.

IV

Thou shalt not mix waste with the ore, neither shalt thou mix ore with the waste, thou nor thy partner, nor the mucker within thy drift, for surely as thou doest these things the mine will stop paying dividends, and thy name will be "mud" over the length and breadth of the camp.

V

Thou shalt not eat onions when going on shift, even though they be as cheap as real estate in Clifton, unless thy partner participateth likewise, for that bulbous root exciteth hard feelings in the heart of the total abstainer, and causeth the interior of a mine to be an unpleasant place.

VI

Thou shalt not address the boss by his Christian name, neither shalt thou contradict him when thou knowest he is lying, but thou shalt meekly say "Yes" or "No" to all that he suggests; and laugh when he laughs and keep on laughing when he relateth a story, even though it be older than thy grandmother.

VII

Thou shalt not steal thy neighbor's mops, nor his picks nor his drills; neither shalt thou carry away on thy person or in thy lunch-basket low-grade ore from the mine, for thou wilt find it will take a lifetime to obtain a mill-run.

VIII

Thou shalt not have an opinion concerning thy place of work, for thy employer payeth a fat salary to a school-of-mines expert for constructing in his mind bonanzas that don't exist, so thou shalt refrain from theorizing, and concentrate thy efforts on drilling and the blasting of an abundance of powder.

IX

Thou shalt not, in order to breathe, steal from the drilling machine compressed air intended for drilling purposes. Thou shalt not go on strike lest thou be turned adrift on a cold and cheerless world; neither shalt thou demand thy pay, for the company's directors in the East know not that thou liveth, neither care they a tinker's dam.

X

Thou shalt work and break ore every day, the Sabbath included, for verily the board of directors aforementioned hath assumed the prerogatives of the Almighty, and if thou refuseth to toil as they dictate thou and thy dog and all that thou possesseth will soon be hitting the trail for Tonopah.[41]

The heaviest burden of the hardrock miner's work, however, was the awful risk of life and limb to which he daily exposed himself. The accident rate in the deep western mines was an order of magnitude higher than in the deep mines of the eastern states and Europe. One out of every thirty western miners was disabled every year in an accident. One out of every eighty was killed. The hardrock miner could thus expect to be either temporarily or permanently disabled in one or more accidents during his lifetime and he had an even chance of being killed in one before he retired.[42]

The preamble of the first western miners' union constitution stressed that "the dangers to which we are continually exposed are, unfortunately, too fully verified by the serious and often fatal accidents that occur in the mines." Indeed the underground miner faced a seemingly endless variety of hazards every time he set foot in the mine. He could be blown to bits in an explosion, drowned in a sump, suffocated and incinerated in a fire, scalded by hot water, crushed by falling rocks or cave-ins, wound up in the machinery, ground under the wheels of an ore car, have his head split open with an ax or pick, his neck broken by being run up into the hoist frame, or his whole body smashed out of shape and reduced to a pulp, or ripped apart and scattered in shreds by falling down a shaft. Moreover new accidents were constantly happening, the likes of which were never before heard of.[43]

Although the hardrock miner spent only a small fraction of his time in the vicinity of shafts and winzes, roughly half of the fatal accidents occurred in them. Dan De Quille, who reported the grisly details of many such falls for the Virginia City *Territorial Enterprise*, concluded that the largest fraction of these accidents resulted from simple absentmindedness on the part of experienced miners. Absentmindedness was something the greenhorn miner never suffered from. Dan noted,

the old miner sometimes forgets where he is, while "where he is" is just what the greenhorn is all the time thinking about. He is always on the lookout for trouble, and he is always holding on to something that has the appearance of being pretty substantial, but a man who has worked in the mines for years will walk into a winze or chute in a musing mood, or run a car into the main shaft and be pulled in after it, which is a thing a green hand has never been known to do.

Much of the miner's apparent absentmindedness, however, may have been drowsiness, faintness, or dizziness caused by foul air. The majority of fatal falls from the cage clearly resulted from dizziness while being hoisted up rapidly out of the hot, steamy levels of the mine into the cold mountain air above. Failure of the hoisting machinery, careless operation by the engineer, and falling tools, timber, and rock also took their toll in the shafts.[44]

A fall down the main shaft was almost certain death. Typical of such tragedies was one that occurred in the Chollar-Potosi mine, where a miner ran an empty car into the shaft and was pulled in after it. He fell eight hundred and ninety feet. Pieces of the shattered car were floating in the sump, but his body sank to the bottom. With grappling hooks his mangled form was finally recovered. His head was sheared off down to the jaw, and both legs and arms were crushed in dozens of places. There was not a whole bone left in his body. In falling he had been dashed against the sides of the shaft and here and there shreds of clothing were found sticking to the timbers. In one place a piece of a sock was found, containing a toe. The pump brought up bloody water for weeks. With black humor Dan De Quille recalled the interruption of the inquest of another such victim, when a man rushed in, with a candle box under his arm, requesting in a reverent tone: "Wait a moment, please—I've got some more of him!"[45]

Even tumbling down an incline in the hot levels of the Comstock could prove fatal, if the scalding water that poured from the rocks had collected in the sump at the bottom. Two men died in the steaming sump of the Hale and Norcross incline in 1877. The first was immersed only to his hips and was pulled out immediately but the skin fell off his limbs from the knees down, and nothing could be done to save him. The second man rolled into the sump and an agonizing death after desperately trying to save himself by clutching at the clay-smeared timber above his head. Marks of his fingers in the slime were seen by the search party, and the ghastly, flesh-cracked corpse was dragged up from the bottom with a pole and hook.[46]

Away from the shaft, caving of overhead rock was the gravest hazard to the miner. Even if the rock had been solid enough when first opened slow fatigue and exposure to air could bring it down later. The clay seams would swell and separate in heavy flakes from the walls and the rotten quartz would crack under the pressure from above until the trembling roof gaped open suddenly and collapsed in a crush of fallen rock.[47]

Most caves involved only a small portion of a level and the man working nearby was more likely to escape with a crushed or broken limb than to be mangled and buried in the mass of rock and splintered timbers. Even if he was caught directly in a cave-in, he was not always killed immediately but might suffer a prolonged torture, while a futile attempt was made to rescue him alive. Although the bodies of most men trapped in cave-ins were ultimately recovered, often it was not until they had been half eaten by rats. Many a simple headboard was erected in the graveyard as a silent indictment of "the criminal carelessness or ignorance of the men who failed to timber and support a mine level properly." Timber, of course, was costly and too often a superintendent found it an easy point of economy.[48]

Blasting accidents were another frequent cause of death or injury, and a missed hole, one that had failed to fire, was a special hazard. A missed hole could not always be detected, for when a round was fired a number of reports might be counted which corresponded with the number of holes "spit," but still there could be a "cut-off" hole still containing sufficient powder to kill half a dozen men. The problem was compounded in larger mines where all the holes were fired between shifts. The men on the next shift had no idea where the previous holes had been set, which had fired or which had not. Thus they might fire a missed hole unexpectedly by starting a new hole nearby or by drilling right into it believing it was an unfinished hole from the preceding shift.[49]

Even when the miner was aware of a misfire, it still presented a danger. On one occasion a hole had misfired and the next shift opened and recharged it. After lighting the fuse and waiting the usual time, one of the miners went back to check. He found the cartridge was "boiling," making what they called a "stinker." He started shoveling up water from the floor to drown it out, and just as he was stooping for the second shovelful the half-burned charge exploded. Fortunately he escaped with only minor injuries.[50]

But the most surely fatal hazard in a mine was fire. The vast amount of timber used in deep mines offered a forest of fuel to the uncontrollable flames. Dan De Quille wrote:

No premature explosion of blasts, crushing in of timbers, caving of earth or rock—no accident of any kind is so much feared or is more terrible than a great fire in a large mine. It is a hell, and often a hell that contains living, moving, breathing, and suffering human beings— not the ethereal and intangible souls of men. . . . A great fire on the surface of the earth is a grand and fearful spectacle, but a great fire hundreds of feet beneath the surface of the earth is terrible—terrible beyond measure or the power of words to express, when we know that far down underneath the ground, which lies so calmly on all sides, giving forth no sound, are scores of human beings pursued by flames and gases, scorched and panting, fleeing into all manner of nooks and corners, there to meet their death.[51]

So serious was the calamity of such fires that little consolation could be found in their infrequency. The most disastrous in the early history of western mining was the Yellow Jacket fire at Gold Hill which started on April 7, 1869. It took the lives of forty-five miners and although it was finally contained after two months, it continued to burn for over a year. As the 1972 fire in the Coeur d'Alenes still shows so tragically, even a hundred years of safety legislation has done little to lessen the horror of such a holocaust.[52]

The countless hazards faced by the hardrock miner inevitably had a strong effect on his psyche. His most common defense was a cool disdain of danger and the fatalistic attitude of the gambler. When a man faced death daily, he soon came to regard it lightly. For if the fear of death was constantly before him he could neither work nor think effectively. He expected to die "when his time came," but he saw no sense in dying a hundred deaths from fear beforehand. He looked upon the dangers of his work as the risks of the game and he staked his life on the cast, because he felt the odds were still in his favor. Even if he were snatched "out of the jaws of death," he was not necessarily overcome with feelings of thanksgiving, but simply regarded his good fortune as a gambler would his winnings. This same gambling spirit also led him to drop his savings at the gaming tables, in the stock market, or on a grub stake for prospecting.[53]

But even though the miner may have considered the risks underground an acceptable gamble, he demanded that his pay be commensurate with those risks. One of the prime aims of the hardrock miners' unions was the maintainance of a wage "proportionate to the dangerous and laborious nature of his employment." The heightened risk underground was the hardrock miner's principal argument for higher wages than the surface worker's. It was also the sole basis for his union's successful, but controversial, demand for a single minimum wage for every man working underground, whether he was a miner, a mucker, a carman, or a common laborer.[54]

Finally, providing relief from these hazards was the fundamental function of the hardrock miner's unions. Indeed, the bulk of the treasure and energy of the unions was devoted to benefiting the victims of the sickness, injury, and death that were the toll of these hazardous conditions. The unions paid a daily subsistence allowance to sick and injured members, established medical insurance programs and even hospitals, attended to the funeral arrangements of the dead, and provided for their widows and children. To reduce these hazards the unions fought for better health and safety measures: changing rooms to cut the risk of pneumonia, ventilation shafts and blowers to clear the foul air, safety cages to reduce accidents in the shafts, adequate timbering to prevent cave-ins, connection of the underground works with more than one shaft or tunnel for escape in case of fire, and appointment of state and territorial mine inspectors to enforce safety standards. With mixed success such measures were demanded of the mining company management under threat of strike and introduced in the legislature by elected miners.

Many of the hazards faced by the miner were brought about by inexperienced or indifferent management. The mines operated by absentee owners were particularly vulnerable to mismanagement, and as the mines went deeper and outside capital-

ization was needed, this problem became a major one to the
miner as well as to the stockholder. Too often the mining company directors or trustees in New York or London knew nothing
of mining and quite probably had never even seen the mine
they owned. They entrusted the management and development
of the mine to an agent, manager, or superintendent who was
often equally ignorant, with the simple faith "that a man having been a Sunday school teacher, or a most exemplary tradesman, or a needy relative of the president or one of the directors
is sufficient qualification to enable him to manage a mine successfully." Such men were incompetent not only to make the
mine pay its stockholders, but also to judge the necessity of
adequate timbering, proper ventilation, and other safety
measures.[55]

Even when the directors tried to find a practical, experienced
man for the job, they were frequently taken in by some char-
latan, whose only talent was making the most for himself at the
expense of both the stockholders and the miners. Many a lar-
cenous superintendent charged his personal expenses to the
company, embezzled funds, took kickbacks on supplies purchased by the company and on contracts let, and gouged the
miners by forcing them to eat in a boardinghouse and trade in
a "pluck-me" store in which he was a silent partner. One notorious superintendent, Sam T. Curtis, who drew fat salaries
totaling $1,300 a month from four Comstock mining companies,
embezzled $15,000 from one company and built a boarding-
house at its expense which he then gave to his mistress. He, of
course, kept the boardinghouse full by demanding that the company's miners board there or lose their jobs. When the directors
caught up with him they broke up the boardinghouse racket,
but generously agreed to let him remain as superintendent and
pay back the embezzled funds from his salary at a $1,000 a
month—until they found they were losing even more because
he was still embezzling about $1,250 a month from them.[56]

The boardinghouse racket flourished in many camps and contributed a good measure to strained labor-management relations.
So too did the practice of paying the miner in "store orders"
negotiable only in merchandise at a "pluck-me" store in which

the superintendent had an interest. The prices paid by the miner for goods in these stores were greatly in excess of those charged by competitors. In the smaller camps where one company dominated the economy, its superintendent could do pretty much as he pleased. But in the larger camps merchants who did not share the patronage of some mining company officer eagerly joined the miners in protesting these rackets and the courts ultimately declared such payment illegal. One irate Tuscarora miner took more direct action against such abuse by setting off dynamite charges under the beds of the superintendent and his merchant partner, but both escaped serious injury.[57]

Occasionally the directors supported such subsidiary enterprises and established company-owned stores and boardinghouses which their men had to patronize. On more than one occasion the miners struck when the company demanded that even married men board in the company boardinghouse. One enterprising mine operator advocated that the companies also go into the saloon business and hire only hard-drinking miners. He argued,

As a rule hard drinkers are hard workers when sober, and any foreman can cite instances of men who earn every cent they receive—and save nothing. These are the men to have and they should be encouraged to go to your camp. The miner who goes to you and says, "I'd like to lay off for a day or two?" and straightway imbibes freely, will, when he has surfeited with enjoyment, return anxiously to work and can be driven that much the more. Encourage the drinker. The company practically gets his work for his board and overalls.

While this suggestion was made with tongue in cheek, many superintendents did in fact become silent partners in saloons with the same practical result for the miners under them.[58]

A few silver mining companies tried paying their men with fancily printed scrip good only at their store. In one instance there were tragic results. The Ivanpah Consolidated Mill and Mining Company in the Mojave Desert had paid its men in company scrip to reduce costs and support a store run by the nephew of one of the directors. Although the miners grumbled, they accepted the system. But when the company suspended operations and left them with a couple of months' back pay in worthless scrip, they filed an attachment suit and seized the mine and mill. The United States Treasury Department also brought suit because the company had failed to pay a 10 percent tax on the $15,000 worth of scrip it had issued. When the federal agent arrived, the miners refused to give up the property. A scuffle ensued, ending in the death of one miner, John McFarlane, who ironically was the discoverer of the mine.[59]

Other companies hedged on miner's wages by a variety of schemes, in addition to store orders and company scrip. Several paid a quarter of their wages in company stock at par when it fetched but a fraction of that on the exchange. Some paid in devaluated greenbacks and trade dollars. When the price of silver dropped sharply, one company cast their silver into bars, dubbed "bullion checks," which they valued at the old rate and tried to issue as pay to their men. To end such practices, the miners' unions uniformly demanded that wages be "payable in Gold Coin," and struck on a number of occasions to enforce that demand.[60]

Of course even in well-managed mines the best interest of the corporation was not necessarily the best for the miner, and disputes inevitably arose over wages, working hours, use of dynamite, employment of Chinese, hospital fees, and a long list of other issues. All of these problems grew more acute as labor forces grew larger and labor-management relations became less personal with the growing industrialization, capitalization, and absentee ownership of the western mines. With these problems grew the hardrock miner's need for collective action through industrial unions. Thus the industrialization and unionization of the western mines developed together in the latter half of the nineteenth century.

The Comstock Unions

The mining labor movement in the West began with the industrialization of deep mining on the Comstock Lode of Nevada in the early 1860s. The opening of these deep mines, ultimately reaching depths of more than 3,000 feet, required much more extensive capitalization and industrialization than had previously been necessary in western mining. It also required much technological innovation to handle the new problems faced in these greater depths and more complex ores. At the same time, the heightened hazards faced by the miner in these deep mines, together with the impersonalization resulting from larger work forces and absentee ownership, required effective labor organization.[1]

Thus as the Comstock mines would lead the West in mineral production and technological advancement for nearly a quarter century, so too would its miners' unions lead the western mining labor movement. For not only was the movement born there, its form and character were shaped in the struggles there. Its spirit of brotherhood, its benevolent policies, its practical goals and its militant tactics were all mined from the gangue of experience on the lode, and the strong unions that finally emerged became a model and inspiration for those later formed throughout the West. But the Comstock unions had faltering beginnings.

In the summer of 1859 the "blasted blue stuff" that had confounded the miners in Gold Canyon for several years was suddenly found to be fabulously rich in silver. The frantic rush that followed brought thousands of miners over the Sierra from the depressed placer camps of California. The slopes of Sun Mountain above Gold Canyon were soon blanketed with claims. But most were worthless. The paying ore all lay in one great lode—the Comstock—stretching for nearly three miles across the eastern slope. Even along the lode the ore was not uniformly distributed. It was concentrated in less than a dozen rich bonanzas.

A few were exposed at the surface; the others were discovered later by deep exploration. As a result, only a few of the many companies that claimed footage on the lode would share in its fabulous wealth. These rich bonanzas would pay the bulk of the nearly $400 million ultimately taken out of the lode.

The early prosperity of the Comstock was sustained by half a dozen mines: the Ophir, the Gould and Curry, the Savage, the Chollar-Potosi, the Original Gold Hill, and the Yellow Jacket—all in bonanza ore. The notion that these ore bodies were virtually inexhaustible led to gross extravagance and inefficiency in the early operation of the paying mines. Moreover the bonanza ore in such mines stimulated extensive exploration in their less fortunate neighbors. Enormous sums were thus expended with the expectation of finding new bonanzas. Speculators preyed on the excitement with wildcat stock schemes. Even the bonanza mines were victimized by blatant stock manipulations and felonious mismanagement.

But as the ore came out of the mines, the collection of tents and dugouts scattered along the lode soon gave way to substantial wood and brick buildings, and the frontier mining camps of Virginia City and Gold Hill were transformed into urban, industrial communities. At the peak of the excitement in the early 1860s there were more than six thousand men on the Comstock. Of these only about half were engaged in mining, and most of those were more interested in staking a claim of their own than in working for wages. The demand for skilled miners thus forced wages up to $4 or better a day, well above the rate in California. Most of the subsequent strain in labor relations on the Comstock resulted from the miners' determination to maintain this wage after the demand that created it had passed. Nonetheless, much of the successful development of the lode must also be credited to the highly skilled miners attracted by this wage.

It was in the midst of the frenzied boom, however, that the first stirrings of the labor movement began on the Comstock— well before the threat of a wage reduction had appeared. On Saturday May 30, 1863, three to four hundred miners held a mass meeting at Virginia City and resolved to form a Miners' Protective Association. Their purpose in organizing was three-fold. They sought not only "the securance to practical miners of a good remuneration for their toil, and the providing of aid and comfort for them in times of sickness and adversity," but also "the exposure and defeat of speculative plans affecting their interests injuriously." This last was perhaps a curious goal for a labor organization, but it obviously reflected the extent to which many of the miners who had put some of their earnings into mining stocks had been victimized by wildcat speculations and stock manipulations. The membership roll was circulated and the following week, on June 6th, the Association was formally organized. Robert D. Ferguson, later a state assemblyman, was elected president, W. C. Bateman, vice-president, and B. J. Shay, secretary. But as times were still prosperous, the majority of the miners apparently did not yet see the need for organization, particularly since no threat was being made to their position. Thus the Association soon dissolved for lack of interest.[2]

The inflated economy of the Comstock was finally punctured in the spring of 1864. A scare that the mines were exhausted brought stock prices crashing down. Many speculative properties went under and even producing mines like the Gould and Curry, the Ophir, and the Savage plummeted to less than a fifth their value in four months. The miners as stockholders were also hurt badly by the collapse. Thus when the company directors set about to economize by cutting wages, compounding the miners' losses, they became "stubbornly rebellious."[3]

The first to attempt a pay cut were the directors of the Uncle Sam mine in Gold Hill. In March they announced a reduction from $4 to $3.50 a day and put on a new foreman, a Cornishman named John Trembath, to implement the cut. The miners, suspecting that Trembath also had a hand in the decision, reacted swiftly. While he was inspecting one of the lower levels the men jumped him, tied him up "like an Egyptian mummy," lashed him to the cable in the shaft, and hoisted him up with a note reading, "Dump this pile of waste dirt from Cornwall." Although Trembath claimed the cut was not his idea, the directors used him as a scapegoat, firing him and reviving the old wage.[4]

The miners had clearly served notice that they were opposed to a wage cut. But some of the managers must still have read

the failure of the Protective Association as an indication that
most of the men would not organize to resist one. If so they
were badly in error, for their subsequent action triggered the
first miners' strike in the West and rallied broad support for the
mining labor movement. Just a few months after the Uncle Sam
incident, several of the managers in San Francisco secretly
agreed to jointly cut wages to $3.50 a day. The date set for the
cut was August 1st but, seeking the advantage of surprise, the
men were not told until they left work on Saturday July 30th.
This only strengthened their opposition.

The following evening nearly every miner and mill hand in
Gold Hill massed to protest the wage cut. They were addressed
by George B. Johnson, a miner in the Empire mine. He called
for a strike and the crowd roared its support. The Gold Hill
Brass and String Band then sounded up as the men, strung out
in a long procession, marched up to Virginia City to enlist the
support of miners there. The band played "all the defiant airs
in their scorebooks," while the throng joined in with cheers and
shouts of "Four dollars a day!" and "No reduction of wages!"
The miners also fired a salute of "unearthly groans" at the house
of Charles Bonner, the foppish, twenty-nine-year-old, superintendent on the Gould and Curry mine. Bonner, fearing for his
neck, had "skedaddled," to hide in the hot, cramped attic above
the kitchen of a French restaurant. There he nearly died of "heat
and fear" before friends dragged him out by the legs and stashed
him in the cooler sanctuary of Father Monaghan's bedroom.[5]

This first strike was to set the general pattern and style of the
miners' actions in future strikes both on the Comstock and elsewhere in the West. At six o'clock the next morning, August 1st,
several hundred miners gathered at the Imperial mine works in
Gold Hill to demand continuation of the $4 wage. The super-
intendent, however, had already decided to capitulate, posting
a notice that all hands could resume work at the old wage. Representatives of a number of other companies, led by James M.
Day, superintendent of the Yellow Jacket, denounced the attempt to reduce wages. As soon as the brass band arrived the
miners marched off to the other works threatening cuts. They
won similar agreements to maintain the old wage at the Sap-

phire and Comet mills and the Empire mine. At the latter they also learned that a rumor that Johnson had been fired for his speech the previous night was unfounded.

"Forward march to the Gould and Curry!" was then ordered and the crowd headed over the divide to Virginia City. There several hundred more miners joined in, as they paraded triumphantly up the length of C Street and back down B to the Gould and Curry office. The building was locked up, so they moved on to Charley Bonner's house, but he still could not be found. The angry miners returned to the office and broke open the door to find the bookkeeper inside. He promised that wages would not be reduced. But when quizzed as to the present rate, he revealed that many of the miners were already getting only $3.50. The miners demanded that all be paid $4, but the bookkeeper could not promise this. Johnson then suggested that they go to the mine and discharge all who were working for less than $4. On the way they stopped at the Mexican and Ophir mines where wages had also been cut. There were no company officers at either mine so the deputation went in to call out all those working at the reduced rate. The full force was turned out of both mines and they were closed down. When they reached the Gould and Curry mine a "committee" of nearly a hundred brought up all those receiving lower wages, leaving the others on the job. The crowd was about to go on to the company's mill when word came that Bonner was willing to meet with them at the International Hotel. There was considerable debate about whether to go chasing after him again, but they finally agreed to meet with him.

At noon with the band playing and flags flying nearly two thousand miners, some sporting sledges, picks, and shovels, packed C Street in front of the hotel. But Bonner still could not be found. While the miners waited a few local politicians, former San Francisco Judge Tilford, senatorial candidate Charles De Long, and others took the opportunity to address them. From the hotel balcony Tilford delivered an "eloquent and burning" opening speech, which set the tone for those to follow. "To reduce wages," he argued, "is to drive to despair and death the miner and his family. It cannot and must not be. By the law of ancient Rome a convicted traitor was hurled from the Tarpeian rock. Let the man who, in this crisis, advocates a reduction of miners' wages be girded and encircled with burning fagots, and receive the fate of the Roman felon!" Advocates of the wage reduction got the message; as one later remarked "though it was not clearly apparent to everyone present that despair and death were included in the proposed reduction of 50 cents, no one cared to win a girdle of fagots or its Washoe substitute by confuting the assertion."[6]

As Bonner had still failed to show up, when the oratory ended at half past two, a committee was chosen to seek out and confer with him and the officers of the Ophir and Mexican companies. The throng then headed back toward the Gould and Curry mill to close it down before dispersing. That evening the miners convened in the District Court Room. There the committee reported to the jubilant crowd that all three companies had received orders from their trustees to pay all hands $4 a day. Thus the miners won their first strike in the western mines.[7]

The miners' success in this strike owed in large part to the broad support that they had in the community. The press on the Comstock had consistently opposed even the rumor of a wage cut. They viewed such action as a threat to the prosperity of the district and they heartily endorsed the miners' strike to uphold wages. Philip Lynch of the Gold Hill *News* declared, "It needs no argument to prove that four dollars a day is as little as a man can live comfortably upon in this land of enormous prices, and that even at that price it requires health, industry and economy in the miner to make ends meet. It may be possible that the price is more than the mines and mills can afford, but we don't believe it!" With much more rhetoric than force the more radical Virginia City *Union* warned, "The opulent owners of silver mines who deceived themselves by the thought of being able to ignore honest labor will find themselves sadly duped and subject themselves through an avaricious cupidity to great inconvenience and loss by allowing their works to lay idle rather than pay a just and reasonable compensation." The *Territorial Enterprise* brushed aside all argument with the simple assertion that "if mines cannot afford to pay $4 a day to miners, let them lie still;

for they will do the Territory little good and their stockholders less."[8]

Jubilant over their success, the miners met again on August 2d to organize a new union. They were encouraged in this aim by several speakers, among them Judge Tilford and De Long's senatorial rival, William M. Stewart. The new organization was styled "The Miners' League of Storey County." William Woodburn, a twenty-six-year-old Irish emigrant and veteran of nearly a decade in the mining camps of California, was made president pro tem. He later served three terms as Nevada's Congressman. A native son, Patrick Henry Burke, who would remain prominent in the movement for three decades, was made secretary pro tem. In addition a committee was appointed to draft a constitution and bylaws. It was headed by George Johnson of Gold Hill and Michael Lynch of Virginia City, and met in Frank Tilford's office.[9]

The following Saturday, August 6th, the League assembled to adopt the constitution. The Virginia City miners came down to Gold Hill in a torchlight procession, carrying giant transparencies proclaiming: "Gould & Curry Mine, $4 per day." "United we stand, divided we fall." "Equal rights, laws & justice," and "August 1st, 1864: Important era in the history of Nevada Territory." Joining the Gold Hill miners, they massed in front of the Vessey House, as there was no hall large enough to hold them all. The constitution and by-laws was read and adopted by unanimous acclaim. Woodburn was elected president, Johnson vice-president, and Terence Masterson secretary. The meeting broke up about midnight and the crowd returning to Virginia City "made the welkin ring with their shouts."[10]

The "sole purpose" of the League, according to its by-laws, was "keeping wages at a standard of Four Dollars per day." Thus every member pledged on "his word of honor never to work in the County of Storey for less than Four Dollars per day, to be paid in gold or silver coin of the United States." The specification of coin was aimed not at company scrip but at greenbacks which at that time brought only forty cents on the dollar in Virginia City. To secure this standard throughout the Lode, each member also pledged to report to the secretary any men found working for less than $4 a day. This information would be passed to the League's trustees for prompt action and "if any remonstrance they make with the employers paying such reduced wages be disregarded, the President shall be required to call out the entire force of this League." Each member was also asked to help fellow members get jobs in preference to nonmembers.[11]

Although the wage issue was clearly the dominant one in the minds of the League's founders, other benefits were not entirely ignored. The constitution provided that a physician be retained at a fixed monthly salary "to give all necessary medical attention to any member of this League." But at the first monthly meeting it was resolved "to dispense with the appointment of a physician at the present time," and a charitable fund was established instead.[12]

The political turmoil of these Civil War years prompted the miners to guard their association against what the Virginia City *Union* termed "the machinations of designing men, who, if they can, will turn the organization into a political machine . . . for their own personal aggrandizement." The by-laws thus cautioned members to "use great vigilance to prevent it from being converted into a political Society," and to insure that "no political discussion shall ever be maintained at any meeting of this League." The League held to this principle, but that did not prevent others from seeking the votes of the League as a block, as did Yellow Jacket superintendent James M. Day, who ran for Sheriff on his record of opposition to the wage cut.[13]

The success of the Storey county miners in resisting the wage cut prompted similar action at the mills in neighboring counties where much of the Comstock ore was worked. On August 25th fifty Ormsby County millmen led by Joseph Sparrow visited all the mills in Empire City on the Carson River and obtained a $4 minimum wage. About the same time the miners and millmen along the Truckee River organized the Miners' League of Washoe County, headed by W. D. Holland. This association was absorbed into the Storey county league in September.[14]

That summer the first stirring of the mining labor movement even touched the distant camps of Aurora and Austin—a hundred miles to the south and a hundred miles to the east. The owners of the Real del Monte mine at Aurora were party to the attempted wage cut of August 1st, but they too met with prompt resistance. On the evening of July 31st some three-hundred miners armed with picks, drills, sledges, and shovels filled the streets of Aurora protesting the cut. Like their Comstock brothers they were accompanied by a brass band and banners demanding "Four dollars or nothing!" They also received the spirited endorsement of both the camp's newspapers, as well as of the City Marshal and local politicians. The following day they too formed a "miners' protective association." It was apparently styled after that attempted on the Comstock the previous year, but members also signed a pledge not to work in the mines for less than $4 a day nor to permit others to do so. D. B. Grant was elected president, R. E. Thompson vice-president, T. J. Murphy secretary, and J. Hays treasurer. Within a few days the directors of the Real del Monte capitulated and restored the $4 wage. In Austin there was no immediate threat to wages but news of the Comstock organization prompted action. Seventy-five miners met in the District Court Room on September 15th and formed a Lander county League. Its president was B. J. Shay, former secretary of the Virginia City Miners' Protective Association.[15]

The easy success of the Leagues, however, proved to be only an illusion. It was soon obvious that the companies had agreed to the League's demand only to gain time. The directors began quietly, but systematically, to break the League. Whenever a cutback was necessary, it was always a League man who found himself out of a job. And when new men were hired preference was given to non-League men, who would secretly agree to work for less that the $4 minimum. In this way, "an opposition which could not be confronted was thus insidiously sapped."[16]

These tactics were quickly felt by the League and the problem became the first order of business at its first monthly meeting on September 5th. A committee of three was appointed to visit all the mines and mills in the county to find out how many non-League men were at work. At a special meeting two weeks later the committee reported that only a small proportion of the League's members were presently employed. In their place they found many Cornishmen "who are aliens and who come here only to hoard their gains in order to carry them back to their native land." This was sure to win the sympathy of the local merchants who wanted the miners' wages spent locally. But that sympathy alone would not remedy the problem.[17]

Direct action was required and a closed shop seemed the only solution, but the league was in a precarious position to make such a demand. Nonetheless after spirited debate they passed a resolution to be posted at every mine and mill and to be published in the local papers. It read:

NOTICE

The Employees of the Mills and Mines of Storey county are hereby notified that after TUESDAY, the 27th day of September, 1864, no person not a member of the Miners' League of Storey county will be permitted to labor in the said Mills and Mines.
Per order of the Miners' League.

There was a strong division of opinion among the members of the League regarding this resolution and William Woodburn resigned as president, apparently as a protest against it. James T. Parker was elected to replace him.[18]

This action also divided the League's supporters among the press. The Virginia City papers found no enthusiasm for it, charging that the League was "going a little too fast for its own good." But Philip Lynch of the Gold Hill *News* defended the justice of the League's cause and appealed to all miners to join it. He concluded, "We admit that it seems a little hard that employers shall not employ just whoever they please, and do just what they please on their own property—but this is the land of Washoe—*and if labor isn't King here we really do not know who is.*"[19]

The efforts of the League to secure a closed shop quickly proved futile. Although many joined up, a sizable number of non-League miners, supported or perhaps intimidated by their employers, flatly refused to join the League. In addition some of the members who had opposed the policy withdrew from the League in protest. The League found itself powerless to enforce the demand. Thus five days before the deadline, the League trustees disavowed the demand for a closed shop. But it was too late to save the now badly divided League.[20]

Lynch of the Gold Hill *News* also retrenched from his earlier stand, conceding, "When the full scale of prices is paid by the Superintendents, it is certainly all that can be asked of them by any Society. We remarked, day before yesterday, . . . that *Labor is King* in Washoe. We used a figurative expression—and will here amend it. *Labor is King*—provided the King behaves himself and does not trespass upon other vested rights."[21]

The League was broken and the superintendents moved swiftly to hasten its distintegration. The very next day they organized their own forces as the "Citizens' Protective Association . . . for the purpose of resisting the demands and assaults of lawless combinations now existing in the county." Despite the name, membership was restricted to mining company representatives. J. B. Winters, who had replaced pro-League James Day as superintendent of the Yellow Jacket, was elected president and Charles Bonner of the Gould and Curry was vice-president. It was the first of many employer organizations formed in the western mines to fight the growing labor movement. They continued to masquerade as "citizens' " associations for two decades before adopting the more appropriate title of Mine Owners' Associations, but absolute opposition to the mining labor movement was their primary goal throughout.

With pledges of support from the governor, mayor, and sheriff, the Association issued a proclamation declaring,

We recognize no right in any person or persons whomsoever to interfere in the relations existing between ourselves and those employed by us—and . . . in the event of any attempt on the part of any persons to commit the threatened interference with the rights of those in our employment, or to force them into any association against their inclinations and contrary to their privileges as citizens, we will demand of the proper officers of the County and Territory, that they use all lawful means to suppress every attempt at the violation of our rights, and the rights of our employees.[22]

The most effective action of the Association, however, was an unannounced decision to accelerate their previous policy of firing League men and hiring non-League men in their places. Within a week several of the superintendents, led by Bonner, laid off their entire crews and promptly began hiring new men.

It was clear that no League men need apply. The new crews were said to have been made up exclusively of "Cornishmen and Secessionists." The superintendents also used this opportunity to secure even greater wage cuts than they had attempted two months before. Although shaft and wet work still drew the old rate of $4 a day, drift work was cut to $3.50, and car and surface work was cut to $3.[23]

The broken and divided League was powerless to resist. They debated in secret sessions but failed to agree upon a course of action. William Woodburn resumed the presidency of the League in an effort to reunite the factions but to little avail. Membership in the League was now "a manifest black mark" with most employers. George Johnson and others prominent in the League had to leave the district to find work. Only Woodburn, who opened a legal practice not dependent on the goodwill of the mine owners, was able to remain. As winter drew closer desertions from the League were legion. By the end of the year it was disbanded. With the breakup of the Storey County League, those in outlying districts were also abandoned. The following spring without the organized opposition of their employees, the superintendents quietly reduced all remaining underground wages to $3.50 a day.[24]

Thus the miners' second attempt at organization also failed. But this time a strong union spirit had been kindled among the miners that would never be quenched. For the first time a majority of the miners had been organized and community support had been rallied in opposition to a wage cut. It was only because a sizable fraction of the miners did not yet appreciate the need for a closed shop to maintain solidarity that the League was finally broken. Advocates of the closed shop learned from the defeat that the only practical course at least for the time being was to try to win this end not as a compulsory demand but as an act of moral responsibility on the part of every miner. Adopting this course, they could hold community support and still fight company blacklisting tactics by demanding only that the companies not discriminate against union men.

Despite continually rising bullion output, the market value

of Comstock mining stocks continued to decline until the end of 1865. By then the mines were clearly underpriced and a gradual recovery began. This was bolstered by the discovery of new, deeper ore bodies which raised the output of the mines to roughly $1 million a month. As prosperity returned, the demand for more miners forced wages back up, and by late 1866 many of the mines were again paying $4 a day.

Encouraged by this revival, the Gold Hill miners once again organized to secure a $4 minimum wage. Learning from their earlier defeats, this time they would succeed even beyond their immediate goals to build a union that would be a model and inspiration to all those that followed. This time too the Cornish joined enthusiastically in the movement. On Saturday evening, December 8, 1866, they met with "Open Doors" in McCluskey's Theater Hall to form "The Miners' Union of the Town of Gold Hill." The driving force in organizing the new union was John G. White, a forty-seven-year-old Cornishman. Jack had come to California in 1852, working in the mines at Mariposa and Grass Valley, and then to the Comstock in 1860. He and others prominent in the new movement were more practical men than the leaders of the old League, but they were also much more militant.[25]

This new spirit is best voiced in the Introduction to the union's constitution, which White and others drafted:

Whereas, in view of the existing evils which the Miners have to endure from the tyrannical oppressive power of Capital, it has become necessary to protest, and to elevate our social condition and maintain a position in society, and that we should cultivate an acquaintance with our fellows in order that we may be the better enabled to form an undivided opposition to acts of 'tyranny'—Therefore, We the Miners of Gold Hill have resolved to form an association for the promotion and protection of our common interests, and to adopt a constitution for its guidance, for without Union we are powerless, with it we are powerful;—and there is no power that can be wielded by Capital or position but which we may boldly defy,—For united we possess strength; let us then act justly and fear not.[26]

But realizing that so militant a declaration would alienate moderate elements in the community, the union replaced this introduction, in its published copies of the Constitution, with a preamble less offensive to capital. The preamble stressed only the perils of the trade and the union's benevolent goals.

WHEREAS, Experience has taught us (the Miners of Gold Hill) that the dangers to which we are continually exposed are, unfortunately, too fully verified by the serious and often fatal accidents that occur in the Mines, and that Benefits in many of these cases are positively necessary; and whereas, "the Laborer is worthy of his hire," proportionate to the dangerous and laborious nature of his employment; and that it is as much to the interest of the Stockholders in mines on the Comstock to retain experienced labor as for the Miners to receive their just recompense; and whereas, an organization, having for its objects the maintenance of these principles and the dissemination of the knowledge of "practical experience" in mining, is not only a benefit to its members, but to the community generally; therefore, that these ends may be attained, we have resolved to form an Association having for its motto: "Justice to all—live and let live"—and pledge ourselves to be governed by the following Constitution and By-Laws.[27]

The militant introduction, however, was still retained in the manuscript version which every new member swore to and signed. Similarly the published aims of the Union were only broadly defined as "the elevation of the Position and Maintainance of the rights of the Miner"—making no specific reference to wages. But the wage issue was fundamental to the organization, and every member also swore to a secret Pledge, promising that he "will not under any Circumstances hereafter work underground in any Mine, for any Company, individual or individuals for a less compensation than Four Dollars ($4.00) per day—payable in Gold Coin."[28]

The new union, like the former League, was essentially an industrial union, rather than a trade union. Although surface workers were excluded, its membership was open to all underground workers—miners, muckers, carmen, pick carriers, and watermen alike—regardless of the level of skill required in their work. The underground workers made up roughly three-fourths of the mining labor force. This concept of industrial unionism, although slightly restricted, became a fundamental principle of the early western mining labor movement and later evolved into the broader concepts of industrial unionism embodied in the Western Federation of Miners.[29]

The constitution adopted by the Gold Hill union proved to be a very workable and durable document. With only slight modification it was later adopted by miners' unions throughout the West and it continued in use for over half a century in all the locals of the Western Federation. Because it was, in fact, so workable, it seems very likely that it was rather closely modeled after that of some other organization. But what that organization might have been remains a mystery. It bore no resemblance to that of the earlier League, or to that of the only other local labor organization, the Washoe Typographical Union formed in June of 1863. Furthermore it is unlikely that Jack White, who took the lead in organizing the union, had any previous experience in the labor movement. He had come to the West straight from Cornwall and the miners there were an enigma among British miners for their "absolute" lack of unionism. Indeed they were often hired as strike breakers in the unionized coalfields. The most likely possibility, therefore, seems to be that the Gold Hill constitution was modeled after one of the coal miners' unions' in the eastern United States, where some of the Comstock miners had worked. The early eastern coal miners' unions did, in fact, embrace the same concept of industrial unionism as the hardrock miners in the West. If this is the case, however, it is particularly ironic that the hardrock miners' unions never worked, or associated in any way, with the coal miners' unions which were also forming in the West at the same time.[30]

Jack White was empowered by the open convention on December 8th to appoint twenty charter members and form the Union. They met two days later and elected temporary officers: William I. Cummings, president; Patrick Neville, vice-president; Angus C. Hay, recording secretary; John G. White, financial secretary; A. A. Putnam, treasurer; John B. Stephens, warden; and C. Albert Prince, conductor. All but Neville had worked on the Comstock for several years and doubtless had been members of the former League. At the first regular meeting on December 15th sixty-one new members were enrolled and during the next two weeks frequent meetings were called to enroll more members. By December 27th, when the first regular election was held, the union numbered 158. All the temporary officers were reelected except Stephens. The union grew to over three hundred men within the next six months, and by the fall of 1867 it held nearly all of the four to five hundred miners in the town.[31]

Within a month of its organization the union launched a campaign to establish the $4 wage for all underground work in Gold Hill. On January 3, 1867, a three-man Committee of Inspection was appointed to determine how many miners were working in each mine for less than $4 a day. They reported that the Imperial, Empire, Logan and Holmes, and Crown Point mines were paying some underground men less than the standard. All the superintendents except J. D. Greentree of the Imperial, however, had treated the committee "politely and courteously" and had led them to believe that "the matter could be adjusted." Greentree, a brusk, former sea captain, had simply told them to "go to hell and mind their own business."[32]

After a few weeks of debate on the best course of action, the union resolved to send a new Committee of Three to notify the superintendents that "unless $4.00 per day are paid to all miners working underground in their mines, the Miners' Association will turn out in force and stop their works until the same is accomplished." When the companies failed to reply to the demand, the union voted to go "in force" to the Imperial mine and shut it down unless Greentree agreed to pay $4. Led by Cummings, Hay and White, nearly two hundred union miners and many onlookers massed at the Imperial hoisting works shortly after noon on Monday February 11th. Although Greentree was away at the time, they called his men out of the mine and by "determined persuasion" discharged several who were working for less than the "union price." The miners then moved on to the Challenge works where they recruited twenty-three new members and called a special meeting at McCluskey's Hall to enroll them.[33]

When Greentree returned to the mine later that afternoon he found a note from the union warning that if he put any men to work underground at less than $4, he did so "at his peril." In response, he promptly discharged three known union men and persuaded the superintendents of the Yellow Jacket, Kentuck, and Crown Point to fire Cummings, Hay, and two other union members. Greentree, however, had acted without the approval of the company's officers in San Francisco. Frantic telegrams announcing the closing of the mine and wild rumors of "blood, massacre, and pillage," caused a frenzy of excitement on the San Francisco Stock Exchange. Imperial stock dropped $10 a share and the company president, William B. Bourn, quickly departed for the mines.[34]

Through attorney Charles De Long, Bourn notified the union of his desire to adjust the matter "amicably," and he agreed to meet with Cummings and Hay on February 27th. Hay made a transcript of this first labor negotiation, which gives some insight into management attitudes at the time.[35]

Bourn first asked why only the Imperial had been struck when other mines were also paying some of their men less than $4 a day. Was it an "antipathy" against the mine or himself? Cummings assured him that there was no enmity against the mine or against him. In fact, he doubted that there were six men in the union who even knew Bourn's name, adding "I did not know it myself until today." He pointed out that since all the officers of the company lived in San Francisco, the superintendent was the only representative the union could reach, and they had concluded to take action against the Imperial because Greentree had refused to talk with their committee and had treated them "very disrespectfully." Bourn suggested that Greentree was just in an "ill humour," and he assured Cummings that "I have seen Mr. Greentree and told him that I regretted very much that he did not treat your Committee with respect." Bourn asked, however, that all future grievances be presented to their attorney, De Long. "He will communicate with us and, so long as your demand is reasonable, rest assured it shall be granted. We shall pay as high wages as is paid in the mines. I am in favor of proper recompense, but do not ask of us what you will not of another."

Bourn did, however, challenge the basic concept of a minimum wage in view of variable abilities and circumstances of individual miners. "It does appear to me that you should make some allowance for different men. You will agree with me, I think, that one man can do more than another. Now in working in hard rock, here is a man strong and robust who can do a good day's work, while another weak man beside could not pick half so much in the day. Is it fair to pay the weak man $4.00 per day as well as the man who did twice as much? I do think your organization should make some distinction there. Again a man may happen along who has a sick sister or mother to support, and still may not be so apt as a practical miner. Should there be no distinction there? Give me your opinion?"

"In the case of the weak man, he has no business in the mine," Cummings responded. "If he can not earn $4.00 per day discharge him; it is not his place. There is plenty of work on top for him. And for the man who may happen to have a sick mother or sister to support, if the latter be a member of our Union, his case will not pass unnoticed."

"Then I see you claim it as a craft," Bourn said, "and if a man can not earn $4.00 per day discharge him; give him work on top. Do you propose to fix the price of labor of men working on top, say on croppings?"

"We will not interfere with them and have no request with regard to them," Cummings replied. "Only those who work underground must have $4.00 per day and no distinction."

"Now if this were known before, all could have been avoided," Bourn concluded.

Cummings then summarized the union's position. "What we want is $4.00 per day for men working under ground, and that no distinction be made whether they be members of the union or not." Bourn agreed to both.

"That is all we ask, and I shall be much pleased to carry the news of your courtesy and explanation to the Union," Cummings replied. In parting he promised Bourn that no Union member would harm Greentree, and assured him none too subtly that "the Military Co's, Fire Co's, etc. of Gold Hill and vicinity are composed principally of miners and they will ever be ready to protect the property and officials of mines paying $4.00 per day."[36]

The union thus obtained recognition of its representation of the miners and acceptance of its basic demands for a minimum wage and no discrimination against union members. Moreover this first meeting between the union and the management had achieved a candid and cordial atmosphere that would characterize nearly all labor negotiations on the Comstock for the next half century.

Following the acceptance of the union demands by the Imperial, the other Gold Hill mines paying less than $4 a day soon brought their wages into line; the Logan and Holmes within days, the Empire by the end of March, and the Crown Point by the end of April. Cummings and other union members also found no further trouble getting work.[37]

The success of the Gold Hill Miners' Union soon led to the formation of a similar union in Virginia City. Jack White helped to set it up and on July 4, 1867, the entire Gold Hill union marched up to Virginia City to take part in the organizing ceremonies in the District Court Room. Sixty-five charter members were appointed, including William Woodburn, Michael C. Lynch, and many other members of the former Miners' League. William Livingston was elected president.[38]

The Virginia City Miners' Union was of course closely modeled after that of Gold Hill and its constitution and bylaws were adopted almost verbatim from those of the Gold Hill union. The two constitutions differ significantly only in their published preambles, which were written later as a more moderate substitute for the militant introduction. The preamble of the Virginia City union like that of the Gold Hill union, emphasized the hazards of the trade and the benevolent aims of the union, but it still retained some of the phrasing of the secret introduction.

WHEREAS, In view of the fearfully-hazardous nature of our vocation, the premature old age and many ills, the result of our unnatural toil; and whereas, a Society, which would enable the Miner to be his own benefactor, would also relieve the Corporations on the Comstock; and, whereas, it is profitable to retain skilled and experienced labor when its demand is significant in proportion to the benefit to be derived from it; and, whereas, we should cultivate an acquaintance with our fellows, in order that we may be better enabled to form an undivided opposition to acts of injustice: Therefore, we, the Miners of Virginia, have resolved to form an Association for the promotion and protection of our common interests, and have adopted the annexed Constitution and By-Laws for its guidance; for united we possess strength. Let us then, act justly and fear not.[39]

Despite the close ties between the two unions, however, the Gold Hill union voted early in July not to allow Virginia City miners to visit their meetings until "they show by their actions that they will strike for their rights." The Virginia City miners responded quickly. An investigating committee learned that the Savage Mining Company was employing some men underground for less than $4 a day. Since its superintendent, Charles Bonner, had played a prominent, if unheroic, role in breaking the former League, the Savage was a particularly attractive target for the new union's first show of strength. Thus on Sunday evening, August 4th, while Bonner was relaxing at Lake Tahoe escaping the heat, the Virginia City miners voted to strike the Savage. Early the following morning several hundred miners, led by Jack White and accompanied by the sheriff and the police chief, marched to the hoisting works and presented their demand for a minimum wage to the foreman, Samuel B. Ferguson. He agreed to hoist out all the men and White questioned each as to his wages. Of the seventy-three men working underground, fourteen were receiving less than $4 a day. White explained to them that "every mine on the Comstock was paying, and willing to pay, $4 per day, that the Committee was there to raise their wages, and that none but men who were receiving full pay would be permitted to go back into the mine."

At this point an embarassed Bonner rushed onto the scene. He announced that as of the first of August every miner working on the breast was receiving the minimum wage, but that carmen, shovelers or muckers, pick carriers, watermen, and outside men were only paid $3.50 a day. The union men objected and after a brief but heated discussion Bonner also agreed to pay carmen and shovelers $4 a day. The miners then gave three and a tiger to both Ferguson and William J. Forbes, editor of the

Daily Trespass, who in turn complimented them on the "quiet-ness and order" of the proceedings.[40]

The Virginia City union had thus shown its ability to establish the minimum wage on the upper end of the Lode, as the Gold Hill union had done on the lower. Demanding only the minimum wage and no discrimination against union members, the two unions gained a broad and secure base of support among both the working miners and the community at large. Thereafter they represented so formidable a force that the mine management uttered only chronic complaints against the "arbitrary standard," but never mustered effective opposition to it.

Although the wage issue was the first order of business for the unions, most of their time and resources were devoted to benevolent functions: aiding the sick and injured, and burying the dead. The benefits to be allotted by the union were not specified in the original constitution and bylaws, and the policy was thus gradually developed as the occasion demanded and the finances of the union allowed. These ad hoc policies were later codified as amendments to the bylaws. The original bylaws did specify, however, that a Visiting Committee of three be appointed monthly to determine the needs of any ailing members.

At first the Gold Hill union tried to pay all the medical and other expenses of the sick or injured, and the burial expenses of the deceased. In addition, each miner usually donated a day's pay to the widow and family. Sick benefits ran as high as $50 and death benefits could go much higher—on one occasion $98 in funeral costs, $100 in medical bills, and $12 for crape arm-bands and publication of a memorial resolution in the Gold Hill *News.* The money for these benefits was drawn principally from monthly dues of 25¢, supplemented by occasional assessments and sundry fines for "intoxication or use of profane language" at a meeting. These limited revenues proved insufficient to provide the benefits required. Thus the union voted on September 19, 1867, to raise the monthly dues to 50¢ and that same month they gave a benefit ball which netted an additional $300. But increased illness that winter again exhausted the union treasury, so it became obvious that some limit must be placed on the benefits. On January 9, 1868, three new sections were added to the bylaws, setting the sick benefit at $10 a week and the death benefit at $100. Within a few years, as medical expenses declined, the sick benefit was reduced to $8 a week. This finally stabilized the union budget.[41]

The monthly income of about $250 from dues of roughly five hundred members balanced average monthly expenditures in sick benefits for four to six men a week and a funeral about once every two months. The benefits worked out by the Gold Hill union also were adopted by the Virginia City union. Final modification of the benefit plan was made by a special committee on October 15, 1877, allowing payment of the sick benefit only after the second week of illness and limiting the number of payments to ten weeks per year. During the next half century the Virginia City union paid out roughly $100,000 in benefits and the Gold Hill union nearly the same. This system of benefits, worked out and refined in the crucible of the Comstock, was later adopted by the miners' unions throughout the West during the latter half of the nineteenth century.[42]

The Comstock unions also subscribed generously to the construction fund for St. Mary's Hospital which opened on March 6, 1876. It was built in Virginia City at cost of $40,000 on land donated by Mrs. Mackay, wife of the bonanza king. Starting in August of 1877 the hospital offered the miners a voluntary medical plan, which provided all medical attention, medicine, and hospital care for $1 a month. The plan had the backing of several mining companies which deducted the monthly fee for the hospital. Within a month there were 120 subscribers, but the majority of the miners preferred to rely solely on their union benefits.[43]

The Comstock miners drew social and intellectual, as well as economic, benefits from their unions. The union hall was in fact the center of social and intellectual life for most of them. Here, in addition to the comradery of weekly meetings, they could enjoy an occasional ball, theatrical performance, or musical and literary society entertainment; while away their evenings at cards, chess, or billiards; or catch up on their reading.

The unions even had complete sets of prospecting gear for loan to their members.

The first miners' union hall in the West, built by the Gold Hill miners, stood for nearly a century. In April of 1868 the Miners' Union Hall Association was incorporated, selling $10 shares to union members and merchants. The following month they purchased a 41-by-100-foot lot on Main Street for $1,200. Construction was slow but the two-story, stone and brick hall was finally completed in the spring of 1870 at a cost of about $4,400. The hall was dedicated with a grand ball and thereafter it rivaled McCluskey's Theater Hall as the social center of Gold Hill. Soon the Virginia City miners built their first hall, but it was destroyed by the terrible fire that devastated much of that town on October 26, 1875. On its ruins they erected a new $15,000 two-story, brick hall which still stands today, as the oldest surviving miners' union hall in the West. It has a large meeting room, which served for their own meetings as well as those of the Mechanics' Union, the Ancient Order of Hibernians, the Knights of the Red Branch, and the Montgomery Guards. In addition it boasted a grand ballroom, a game room, and a library.[44]

The Miners' Union Library was, in fact, the only public library on the Comstock. It was established on December 28, 1877, with $2,000 from the union treasury. The editor of the *Territorial Enterprise* praised it from the start. "Every book added to this library means a further and further withdrawal from saloons and places where men lose money at cards. The library means a lifting up of men's minds and a softening of their hearts, and is more needed in Virginia than is either food or clothes." By 1881 the library held 2,200 volumes and ten years later 4,015 volumes and 26 newspaper and magazine subscriptions. For many years it was the largest library in the state. The books were selected by a board of five directors elected by the union. They bought only those books that would be widely read—novels, romances, travels, and elementary texts on mechanics and physics. As a result the books were thumb-worn and frequently soiled, but none were dusty or had uncut pages.

Union members had free use of the library; nonmembers paid 50¢ a month. The duties of librarian fell to the union's financial secretary.[45]

The most important events on the miner's social calendar were the annual miners' union picnic, its anniversary ball and banquet, and the Fourth of July. The annual picnic and train excursion was a joint celebration of the Virginia City, Gold Hill, and Silver City unions begun in the fall of 1874. All the mines on the Comstock closed down for the day while five thousand miners, wives, and children frolicked at Treadway's Ranch on the bank of the Carson River. The anniversary balls and banquets, though formal, were also well attended, fund-raising affairs to which the public was invited. In Virginia City the Fourth of July was also the anniversary of the founding of the miners' union. The miners usually organized the events of the day and always turned out in full force, making up the bulk of the parade and enlivening the festivities with drilling and mucking contests.[46]

The miners' unions also took an active part in local and state politics. Unlike the former League, the unions made no laws banning political discussion at meetings. They had come to appreciate the need for political action both as a possible means of improving working conditions and, more important, as the only means of combating legislation that would threaten the operation, even the existence, of the unions. In practice the miners proved to be much more effective in blocking legislation they opposed than in securing the passage of what they supported.

The first legal threat to the union movement came in January of 1869 as the outgrowth of the expulsion of Chinese laborers from John Fall's mill at the fading camp of Unionville, over a hundred miles northeast of the Comstock.[47] Within a week of the incident the local assemblyman, J. M. Woodworth, backed by Fall and other mine owners, introduced a legislative bill that was not only broad enough to prohibit such specific acts of force as the Chinese expulsion but also the militant strike tactics of the Comstock unions. Entitled "An Act for the Protection of Labor," the bill made it a felony for two or more persons to prevent "by force or show of force, or by threats or intimidations" any person from engaging in, or employing others in, any occupation "on such terms as he may choose." It made equally guilty any person present for the purpose of "encouraging" such an act. The county sheriff was furthermore empowered to hire the necessary force at $5 a day plus board to protect any employee, or employer, so threatened.

The strikes by which the unions had achieved the minimum wage would clearly have been in violation of such a law and the unions responded promptly. Within days a joint committee of the Gold Hill and Virginia City unions published a "solemn protest against the introduction and passage of a measure so directly in conflict with our best interests and the pride of our race." The bulk of their argument, however, was a racist diatribe against Chinese "peonage and paganism." The Committee then circulated a more moderate remonstrance emphasizing, if somewhat abstrusely, that the bill "contemplates a vital encroachment upon their labor-protecting societies." The remonstrance gained wide support, collecting 2,700 signatures, and was presented to the legislature. The bill was tabled.[48]

During the same session of the legislature two bills were introduced to limit the working day to eight hours and to "provide greater protection for the lives of miners." The unions however were unable to muster the same support for these measures as they had in opposition to the previous bill and neither passed.[49]

At the next election in November of 1870, the unions entered candidates of their own on the Republican ticket. James Phelan, president of the Virginia City union, was elected to the State Senate, and Angus Hay and George W. Rogers, both of the Gold Hill union, won seats in the Assembly. Although Phelan and Hay served on the Committee of Mines and Mining Interests in their respective houses, they failed to produce any legislation of value to the miners. In subsequent elections union men also won seats in the legislature. But they faced strong opposition from the owners if they attempted to introduce anything in the way of labor legislation. Nonetheless, union men did eventually help secure passage of a mechanic's lien law and a law exempting miners' wages from attachment.[50]

On the local level the unions held much stronger control. The Gold Hill union had moved quickly to secure sympathetic law enforcement. In June of 1867 the union's first president, William I. Cummings, won an easy victory in a three-way race for the office of town marshal. On assuming office Cummings resigned the presidency of the union but not his membership. The union presented him with an engraved gold and silver marshal's star. The following year Cummings was elected sheriff of Storey County and he served his fellow miners well, if not the cause of justice, in his "feeble protest" against their raid on Chinese railroad laborers in the fall of 1869. For the next decade the office of president of the Gold Hill union was the key step to election as sheriff of Storey County. Cummings was succeeded in turn by Thomas A. Atkinson and Thomas E. Kelly, both past presidents and charter members of the union. In 1878 the office finally passed to a member of the Virginia City union.[51]

The control of local law enforcement by the miners' unions, though rarely used directly, significantly strengthened the unions' position and tended to counterbalance the control that the large mining companies exerted over the state government and militia. Had the companies in fact controlled both local and state forces, they would doubtless have been more willing to risk a strike to win wage cuts, as they could have readily broken it through legal intervention. Similar control was sought by the unions elsewhere in the West, but they were not always as successful as the Comstock unions.

On one occasion the Comstock unions took the law enforcement directly into their own hands when an armed dispute between two mining companies threatened the lives of the men underground. In December of 1877 the Alta Mining Company struck an ore body on ground adjoining that of the Justice company, and the latter promptly claimed the ore. A confusion of suits, countersuits, fraudulent quitclaims, and stock manipulations followed as both companies began running drifts into the ore. By mid-January the drifts were close enough that they could hear the sounds of each other's picks. A connection between the two drifts was imminent and each company hired "a party of professional fighters, armed with shotguns and revolvers, ready to commence hostilities as soon as the connection was made." These "shotgun miners" were paid $20 a day.[52]

Sam Curtis, the superintendent of the Justice, made elaborate preparations to invade the Alta. He built a bullet-proof car with portholes to shoot through. With his gunmen in this "monitor" he planned to capture the drift and if possible the entire level. To smoke out the Alta men in advance of the assault, he also brought down a carload of stinkpots. Charles Derby of the Alta responded with a bulkhead and a blower.[53]

The sheriff got wind of the affair, but he could take no action until a fight started. Several men had been killed in a similar dispute three years before, however, so this time the miners were determined to put an end to the affair before there was bloodshed. The Gold Hill Miners' Union first tried asking that Derby and Curtis take the gunmen out of the mines, simply because they were nonunion men and they had agreed only to work union men underground. Derby consented if Curtis would also agree. But Curtis refused, claiming the gunmen were only friends visiting the mines and not working in them.[54]

The local press then joined the miners in calling for the removal of the gunmen. The Gold Hill *News* declared that "all cool-headed, squarely-reflecting, justly-minded citizens" condemned the action of the two companies and the *Territorial Enterprise* suggested the union take "a retaliation which will stop such outrages on the Comstock for all time to come."[55]

Before the union took action, however, they wanted the support, or at least the acquiescence, of most of the Comstock superintendents. After some discussion a majority agreed to support whatever course the union chose to take, if Derby and Curtis did not remove their gunmen at once. Derby again agreed if all work was stopped in the disputed ground, but Curtis vowed to continue work "at all hazards, and those who interfered must take the consequences."[56]

Thus on January 17th, several hundred miners met in the Union Hall to decide upon a course of action. Although some felt that they should not interfere, most felt that the union had a right to protect its members and since "the expense of the funeral and relief of the family of a dead miner, fell upon the

members of the union, it was a fair proposition for them to
prevent funerals if they could." They voted, therefore, to go
immediately to the mines, take out the fighters, and, if necessary, take possession of the works. Many joined the march as
they moved down the canyon and others lined the road to cheer
them on. Their victory was swift. They hoisted out the "shotgun miners" from the Alta first and by the time they reached
the Justice, Curtis had discharged his gunmen too. Union men
were left as "watchers" at both mines to ensure that the gunmen did not return. The companies paid them $4 a day for their
trouble.[57]

The press praised the union for their "peaceful victory." The
Chronicle commended them for their intervention "in the interest of peace and humanity where the costly and complicated
machinery of the law had been powerless to afford relief." And
the *Enterprise* enthused, "such exhibitions of their strength in
the cause of justice and mercy are in obedience to the highest
laws known to humanity." The course of action adopted by the
union in this dispute was later followed by other unions in
similar instances elsewhere in the West, but they did not always
receive such unbridled praise for their actions. They did, however, usually stop further bloodshed.[58]

The Comstock miners' unions prospered with the development of the Lode and the discovery of new bonanzas. Their
numerical strength reached more than three thousand at the
peak of work in the big bonanza in 1876–77. During this time
they also encouraged and supported the organization of other
workers on the Comstock. The Mechanics' Union of Storey
County was formed in December of 1877 with the aid of the
miners' unions. Sharing the miners' broader philosophy of unionism, it too was not a strict trade union, its membership including
a wide variety of workers in addition to mechanics. One critic
termed it a "conglomerate mass . . . composed of twenty-nine
different crafts; from and between the scientific machinist to the
jack carpenters, are to be found tinkers, tailors, printers and
barbers." This broad base gave them much more strength. The
Mechanics' Union of Lyon County and the Stationary Engineers

of Nevada were organized soon after. In 1878 these unions, backed by the miners' unions, established eight-hour shifts and a $5-a-day minimum wage for the mechanics and engineers working in the mines and mills.[50]

Although the miners' unions forcefully resisted all attempts at cheapening wages, they wholeheartedly supported technological innovations that allowed the companies to make cheap servants of dynamite, steam, and compressed air. In 1869 the miners even went as far as to give financial aid to Adolph Sutro when he was unable to find backing for his projected tunnel to drain the Comstock Lode. The tunnel could substantially reduce costs for many of the mines, draining much of the Lode without pumping and removing ore without hoisting. But William Sharon, Darius Mills, and the rest of the "mill ring" saw it as a threat to their monopoly on hauling and milling the Comstock ores, so they did all in their power to block the project. The tragic Yellow Jacket fire, however, spurred the miners to action. On August 25, 1869, the Virginia City and Gold Hill miners' unions met in joint session and unanimously resolved to subscribe to $50,000 worth of stock in the Sutro Tunnel Company, payable immediately so that work could commence without delay. They declared,

we, the miners, who are compelled to delve and toil daily in an atmosphere heated and corrupted to such a degree that our health becomes impaired, in many instances resulting in consumption and an early death, *are the parties* most deeply interested in the construction of this great work, which had it been in existence at the late diastrous fire . . . which hurled into eternity forty-two of our brethren, would have given them the means of egress, and thus saved their lives.

The official groundbreaking was held two months later with a grand barbecue. The miners' unions' support was the turning point in Sutro's fight for financial backing. Armed with this "home endorsement" he persuaded British investors to back the project. For their part the unions allowed miners working on the tunnel to take $1 of their day's pay in stock. Work on the tunnel, ultimately employing about four hundred miners, led to the organization of a third Comstock union at Silver City, near the tunnel entrance, on March 14, 1874. The tunnel was finally completed in the summer of 1878 and did much to sustain the mining industry on the lode after the big bonanza was exhausted.[60]

Considering the fatal division that the former miners' league had suffered over the closed-shop issue, it is not surprising that the miners' unions were slow in raising the issue again. When they finally did in the summer of 1877, it was only a perfunctory demand for the management's recognition of an accomplished fact, for by then nearly all the miners on the Comstock already belonged to the miners' union. Early in August the Gold Hill, Virginia City, and Silver City unions issued a joint "Notice to Miners of Storey and Lyon Counties," announcing that as of September 10th "No man, except superintendents and foremen of the mines, will be permitted to go underground unless he can exhibit his card of regular membership of the Union to which he belongs or may wish to belong." They publicly defended the demand with the simple argument that

the time is past for one set of men to pocket their four dollars per day and never give up a dime of it toward the support of sick and wounded miners and the widows and orphans of those killed, while another set is now being and has been for years heavily taxed for these purposes. If men work in the mines, the idea is that they shall help to support those that are maimed or made destitute by the accidents that occur in the mines among their fellow workmen.[61]

The miners' demand was supported by most of the local press, but a few seemed to panic. Charles Goodwin, editor of Sharon's *Territorial Enterprise*, decried the demand as "uncalled for, arbitrary and un-American," and he fearfully predicted a violent confrontation that "would leave the city a wreck and make it impossible to work the mines for years." The editor of the San Francisco *Stock Exchange* fancied it the first step in a sinister conspiracy, warning, "It means a strike for higher wages and less work. These Unions intend to drive away from the Comstock every miner who does not, or cannot, or will not belong to them. Then they intend to go for $5 a day, and six hours work —four shifts instead of three. That is what it means and nothing else. The plan has been hatching for months, but this is the first open move towards the consummation." Like others friendly to the unions, the *Eureka Sentinel* denounced such charges as "bosh" and expressed the hope that the move would succeed not only on the Comstock but in every mining camp in the state.[62]

The unions immediately began negotiations with the superintendents to win the demand quickly and peacefully. Two weeks before the deadline the superintendents signed an agreement to "discharge from the mines any miner reported to them as not being a member and refusing to become one."[63]

Thus the dreaded day passed without incident, as the Gold Hill *News* scoffed.

Today is the tenth of September, the day on which according to some of our newspaper contemporaries, the miners of the Comstock were to rise in open rebellion, and by force of arms, and all that sort of thing, take possession of the mines and ruthlessly eject and drive out all miners working below ground who did not or would not belong to their organization . . . . The terrible tenth has arrived, and all is quiet as usual along the Comstock. . . .No standing armies with serried rows of bayonets are guarding the various mining works against incendiary onslaughts of ferocious miners, and the only extra commotion is that of the elements. Old Boreas is the only individual who is on the rampage, and as he rides the fiery untamed Washoe Zephyr over the Divide and along the Comstock he laughs derisively at the fears of those alarmists, and smiles on the terrible insurrection which has ended in wind.[64]

The easy success of the unions on this issue was attributed primarily to the influence of James Fair of the Consolidated Virginia and California—the bonanza mines, which employed more than half of the men on the Lode. Once a miner himself, Fair was an outspoken advocate of the $4 wage and miners' rights. But perhaps even more important, since the miners' demand would cost him nothing, he desired a speedy settlement so as not to interrupt work in his mines which reached their peak production that year.

Only once in better than a quarter century after its establishment on the Comstock was the minimum wage ever seriously challenged. The challenge came at a time of economic crisis on the Lode and the union men, appreciating the problem facing the management, wisely sought an alternative solution that still maintained their own position. The first rumblings for the wage reduction began in 1880 after the exhaustion of the Consolidated Virginia and California's big bonanza. In three years bullion production had dropped to a sixth of its former value and the big mines had started calling for "Irish dividends" as they searched deeper for new bonanzas. When the talk of a pay cut was first heard, Sam Davis of the Carson *Appeal* quickly countered that "nobody who ever went down into one of these mines will think $4 per day too much pay for the men working there." And he jibed, "it is thought pretty hard by some stockholders to have to pay what they call high wages, during times when the mines are not paying dividends. It did not occur to these same stockholders to raise wages any higher when the mines were paying dividends, though the work is just as hard one time as another." Still he suggested if economy were necessary, it

ought to be inaugurated at the offices instead of the mines. If employees have to be reduced in number, in a mining company, it seems as if the non-producers should first be dropped. Without the working miner the mine becomes a mere stock-jobbing operation, unworthy of notice. If wages are to be cut down, the people placed in position by the nepotism of directors or officers, the president and the hangers-on should first be looked to. Let the companies begin at the highest, instead of the lowest, in cutting down expenses. Let them cut down salaries of presidents first. They are useless officers, anyhow, and might be entirely dispensed with for that matter. Mines can be run without presidents or any officer at all, but they cannot be run without miners.[65]

There was, however, much lower-grade ore exposed in the upper levels of some of the mines which could not be worked economically with $4-a-day miners. To work these levels a graduated wage scale and the introduction of child labor was suggested in Sharon's *Territorial Enterprise*. The editor argued that men working in the cooler, drier upper levels should not be paid as much as those in the hot, wet levels, and he proposed an elaborate graded scale, paying men above the 500-foot level only $2.50 a day, from there down to the level of the Sutro Tunnel $3, from there to the 2,500-foot level $3.50, and below that the old rate of $4. Further claiming that there were hundreds of boys in the city in danger of becoming "hoodlums," he sug-

gested that they learn a useful trade by working in the upper
levels at $2 a day.[66]

The unions were solidly opposed to both a wage cut and the
introduction of child labor, but rather than simply rejecting the
proposal they offered a constructive counterproposal. Although
every miner had pledged not to work for less than $4 a day, the
union had never tried to regulate how many hours constituted
a day's work. Most of the mines had worked ten-hour shifts until the spring of 1872 when John P. Jones and William Sharon
cut the shifts to eight hours in their mines as a bid for popularity
in their rival senatorial campaigns. The unions therefore suggested that above the 1,500-foot level miners again work ten-
hour shifts at $4. This was an acceptable alternative for the
mining companies. Work in the upper levels began in March
of 1881 and did much to sustain the Comstock mining industry
in the lean years that followed.[67]

Throughout the latter third of the nineteenth century the
Comstock unions were the bastion of unionism in the western
mines. As such they drew enthusiastic praise from their friends
and bitter condemnation from their foes. Former miner Judge
C. E. Mack lauded their "mighty influence in the cause of labor
and the uplift of human character," while journalist Eliot Lord
damned them as "unworthy champions of the labor cause," who
"substitute might for right" and seek self-aggrandizement rather
than self-preservation.[68]

Both friend and foe, however, were intrigued by their remarkable strength. Judge Mack saw "the secret of the success
of these unions" in their "conservatism" in insisting on only two
demands, the minimum wage and the closed shop. Although
these demands were certainly not viewed as conservative by
the mining companies, it might well have been that if the unions
had made broader demands, such as the regulation of hours,
they could have lost community support and even raised division within their own ranks. The unions' demands were conservative in the sense that they sought only to maintain the
status quo—holding the line against a wage reduction rather
than asking for a wage increase. This gave them a clear advantage in appeal for public support since the burden of proof

then lay on the mine managers to show the need for a wage reduction rather than some other form of economy.[69]

Eliot Lord also pondered the question, concluding that the
unions owed their success primarily to the enduring produc-
tivity of the Lode and to the gambling spirit of the owners. But
he also hinted rather pointedly that "if the Union Milling and
Mining Company and the Virginia and Truckee Railroad Company had not been formed, it is most probable that the union
of miners would have melted away in 1870, as the league, its
predecessor, had dissolved in 1865." He chose not to elaborate
on this point, but here in fact may have been the underlying
reason for the unions' seemingly unchallenged strength. These
two corporations, formed in 1867 by Sharon, Mills, and the
"bank ring," held a monopoly on the shipping and milling of
ores and took the largest share of the profits from the Comstock
mines. They took these profits, not on the value of the ore itself,
but from charges of the sheer mass of the ore that they handled.
Since they thus profited whether or not the mining companies
made a profit, the "bank ring" gained control of the director-
ship of the mining companies to insure that the mines kept
extracting ore for them to haul and mill even if it meant running the mines at a loss which the stockholders had to support
by assessments. Thus they were not willing to risk a long strike,
closing the mines and cutting off their own profits, for the
financial betterment of the mining companies from whose profits
they gained very little. Although conditions later changed somewhat, this situation certainly contributed to the early success
of the Comstock unions. Once the miners became aware of the
power they held as a union they were determined to maintain
it. This determination, their numerical strength, and their broad
support in the community were enough to convince the mining
and milling companies that any attempt to break the unions
would prove too costly if not futile.[70]

For whatever the factors the Comstock miners' unions clearly
demonstrated that labor was king on the Comstock, as they
maintained for their members a level of wages and benefits un-
surpassed in the western mines. But their role in the western
mining labor movement extended far beyond the Comstock.

They were not only the initial thrust of the movement, they were its sustaining force for decades. They provided the example of organization and action that subsequent unions emulated and they inspired the miners with a spirit of unionism that carried the movement throughout the West. The Comstock unions became almost legendary in the western mining camps and their influence was strongly felt in the mining labor movement long after the Comstock mines themselves had ceased to be important.

The Turbulent Years

The early success of the Comstock miners' unions spurred the awakening of the mining labor movement throughout the West. Within six years of the establishment of the Comstock unions, miners' unions had arisen with varying success in most of the major deep-mining camps in the West—from Grass Valley, California, to Central City, Colorado, and from Silver City, Idaho, to Pioche in southern Nevada. All but a couple of these unions were organized by former Comstock miners and patterned from the little pocket book editions of the Gold Hill and Virginia City union constitutions.

The spread of the mining labor movement was as turbulent as it was rapid. For in this first half dozen years the movement succeeded both in broadening its base in the West and in expanding its goals well beyond the simple maintenance of a minimum wage. Some of the new unions struck not just against wage reductions but against long working hours, tyrannical foremen, the introduction of dynamite, and the employment of Chinese miners. Frustrated in their efforts to win their demands, some of the miners resorted to threats and assaults, while the managers fought to thwart them by every means at their disposal, from hired gunmen and agents provocateurs to mass arrests and state militia—the first use of troops in a western labor dispute. It was thus a bitter and tiring struggle but ultimately the unions won more than they lost.

The first attempts to revive the movement outside the Comstock, however, failed not so much from opposition of the mine owners, as from the unsettled nature of the camps and the miners themselves. Though Austin was nearly as old a town as Virginia City and Gold Hill, it was a much less stable community. It was the supply camp for prospectors roaming throughout central Nevada and their seemingly endless reports of rich new strikes in the neighboring ranges kept the boom fever alive in the camp. Early in March 1868, however, many of the working miners met to consider the formation of a Miners' League. Two superintendents also attended: Edwin A. Sherman, a former editor and a strong supporter of the miners' protective association at Aurora in 1864, and Dr. John Goodfellow, a British dentist-turned-mining expert. Sherman enthusiastically urged the formation of a league, while Goodfellow was strongly opposed, trying to convince the miners that "a faithful and capable workman would always receive work and good wages, league or no league."[1]

On March 15th the Miners' League of Austin was organized with James W. Brown as president. The Gold Hill Miners' Union sent a letter of congratulations, and the local *Reese River Reveille* and William Forbes's Virginia City *Trespass* both praised the organization. The league might have prospered, had not the rush to White Pine started almost immediately. Austin was the closest camp to the new excitement and with the first news of fabulous discoveries most of her miners, superintendents, and merchants as well, all rushed off to seek their fortunes. Miners became scarce in Austin and those who chose to stay behind did indeed find that they could get good wages, league or no league. But the union movement like the miners also found new ground at White Pine.[2]

For a time the White Pine mines were expected to rival the Comstock and they drew many men from there. The rush brought three thousand men by the winter of 1868–69 and thousands more the following summer. The richest mines, boasting ore averaging well over $100 a ton in silver, were scattered along the crest of Treasure Hill. Hamilton, a mile north of the mines, was the first town in the district and served as its shipping and commercial center and eventually the county seat. Most of the miners lived in Treasure City perched near the mines in a saddle at the top of the hill. Shermantown, laid out by Edwin Sherman in Silver Canyon a mile west of the mines, was the milling center. Living expenses were nearly three times those in more settled camps; board was $18 to $21 a week and lodging was difficult to find at $7 a week. Wages of $5 a day only partly compensated, but few wanted wages, for the hope of staking a rich claim sustained their spirits, if not their stomachs, at least for a time.[3]

The first winter was particularly severe. By January of 1869 several men had frozen to death or died of pneumonia and many more were ill. Jack White, who had come with many others from the Comstock, wrote back to friends, "We are in the midst of the most terrific storm I ever experienced. In the name of common humanity try and induce people not to rush here so madly at the present time. Tell them all not to make fools of themselves, but to stay quietly wherever they are, and control their impatience for a few short months, when the fearful rigors of winter shall have at least partially subsided."[4]

White soon took more direct action to aid the sufferers. On January 9th he helped organize the Miners' Benevolent Association of White Pine for the purpose of "alleviating the wants and anticipating the wishes of the sick and distressed miner." White was elected president. Despite the obvious need for such an association, it still met with some opposition. In exasperation White wrote of the local Judge, "Burns, the Dogberry of this precinct, refused the use for one hour of his 7-by-9 courtroom to the Miners' Benevolent Association on the grounds that they were an illegal body! They asked him what he knew about it to justify him in such an opinion. 'He didn't know; he didn't know anything.' Condemned without hearing!—benevolence a vice, humanity a crime!"[5]

The Benevolent Association had the support of most of the miners for a time, but when a dispute arose over working hours in one of the mines, White, reflecting the Comstock unions' conservatism on the issue of working hours, refused to take any action. As a result the more militant miners bolted the association. They met at O'Connor's saloon on April 21, 1869, to form the Miners' Union of Treasure City. Daniel McMurtry was chosen president; Patrick Henry Burke, vice-president; and P. J. Tiernan, secretary. Burke had been secretary pro tem of the Storey County Miners' League and Tiernan had been prominent in the Virginia City Miners' Union. Ironically the constitution and bylaws of the new union followed closely that which White had helped to draft for the Gold Hill Miners' Union.[6]

The dispute had arisen when the Consolidated Chloride Flat Mining Company's foreman, Samuel B. Ferguson, suspending some underground work, had transferred the miners to the surface to do sorting and other work. Miners in the district were receiving $5 for a ten-hour day, except for those working on the surface in unopened mines where they drew the same pay for eight or even six hours when the weather was too severe for a longer shift. Thus the miners transferred to the surface asked to work only eight hours. When Ferguson refused, they walked off the job. Rumors of a general wage reduction heightened the excitement and Ferguson received an anonymous threat, warning him to leave the district. But the union disavowed the threat and the walkout quickly fizzled out for lack of sufficient support.[7]

The new Union then decided to concentrate all its efforts on winning the allegiance of the miners away from the Benevolent Association. In this it was much more successful. Within two months White's Association virtually dissolved and the Miners' Union became the unchallenged spokesman of the White Pine miners. By summer most of the working miners belonged to the union. Seven hundred and forty strong, they made up the main body of the Fourth of July parade, where *White Pine News* editor and former Comstocker William J. Forbes praised them as "the bone and sinew of White Pine." There were roughly a thousand working miners in the district, about half of whom worked in small mines with crews of less than a dozen. The largest employers were the Consolidated Chloride Flat with 145 men, the Hidden Treasure with 75, the Aurora Consolidated with 65, and the Eberhardt with 60.[8]

The strength of the union, however, was threatened by the large number of unemployed miners in the district. The completion of the transcontinental railroad in the spring had aided the rush, swelling the population at the mines to several thousand. Most of these men, after an unsuccessful stint at prospecting, were now eager for work to earn enough to go elsewhere. The labor surplus, though only temporary, provided an apparently irresistible opportunity for a wage cut.

On July 10th, Lavern Barris, superintendent and part owner of the Eberhardt reduced wages to $4 a day. All the men in the mine quit work and union members vowed to hoist out any man who went to work for less than $5. Many new men turned up the following morning, expressing a willingness to go to work for $4, but feared to do so because of "whispered and half-uttered threats." Barris thus decided not to try to reopen the mine immediately. The union met that evening, appointing a committee to confer with him.[9]

The White Pine press was unanimous at first in its opposition to the reduction and in its praise of the union's "coolness and deliberation." Even the *Inland Empire*'s editor James J. Ayers, a good friend of one of the Eberhardt owners, commiserated with the miners against the owners, charging "The blessings that accident, or chance, or Providence, has showered upon them, have hardened their hearts, and though the dumps of their mines are crowded and the vaults of their treasure chambers glisten with untold millions, they still covet the pittance meted out to you." He speculated that this was but "the first grand maneuvre in a system of tactics intended to drive off and discourage Eastern capital from investing in our mines, and ultimately starving impecunious owners into selling for a trifle." Forbes of the *News* expressed similar feelings that the wage cut was just "a job put up to bear White Pine in the stock market."[10]

The owners of the Eberhardt were unmoved by such criticism, however. Pointing out that their mine was not likely to cave in or fill with water while lying idle, they declared simply to outwait the miners. A few days later Dr. John Goodfellow, who had landed a job as superintendent of the South Aurora and the Aurora Consolidated, joined the move to cut wages to $4 and his crew also quit. Forbes charged that there was a secret agreement that still other mines would join in the cut. The following morning, the union, 340 strong, visited all the mines along the hill to determine what wages were being paid. They found that the only other mine contemplating a reduction was the Treasure

Hill. The rest of the superintendents, employing over four-fifths of the miners, planned to stand by the $5 rate. The miners gave each three cheers and returned to town, confident that they could win the strike.[11]

Ayers did not share their confidence, however, as he warned:

On the Comstock the miners could easily enforce a fixed standard of wages, but here the surroundings are vastly different. The large milling interest of Storey county which must suffer from a suspension of work on the Comstock, coupled with the damage accruing to the mines by their filling with water during any considerable delay in operations, will always serve as a powerful inducement for the mine-owner to accede to the demand of the miner. But miners have no such advantage here, and when the matter of wages simmers down, if unfortunately ever, to a siege, the employer must eventually triumph.[12]

Although time was clearly on their side, the Eberhardt owners decided to take more immediate action to achieve an effective reduction without a direct confrontation. Within a week of the strike they let a contract for the extraction of 20 tons of ore a day at $1 a ton. Since a man could not even sort and sack more than 4 tons of ore a day, Forbes pointed out the simple arithmetic, if a man took the contract and employed four others at $5 a day, he would have to throw in his own labor free to make good the loss on wages. When five other mines also started up on contract work, he and Ayers both warned the miners that a reduction was now inevitable. "Capital is King," Ayers wrote, "and the republic of labor cannot by force tear one jewel from its golden crown." Forbes suggested that the miners compromise at $4.50 a day. There were brief rumors that alternately the Eberhardt owners and the miners' union were agreeable to the compromise, but no agreement resulted.[13]

The union finally concluded to stop all contract work where men were working for less than $5. On July 27th nearly two hundred miners led by union president Peter Leonard again visited the mines. Some of the superintendents, learning in advance of the visit, dismissed their men for the day. But the miners found about a dozen $4 men on contract work at the Eberhardt and three other mines. All were taken to the top of Treasure Hill, warned not to work for less than $5, and released.[14]

Three days later the strikers faced a new confrontation. Sam Ferguson announced that the Consolidated Chloride Flat would join in the reduction on August 2d and that all of his men had agreed to the lower rate. Dr. Goodfellow and a couple of other superintendents also decided to reopen at the lower rate on that date. With the support of the superintendents a "counter league" was formed by the miners willing to work for $4 a day. Its membership was variously reported between 150 and 500 men. The sheriff promised to be on hand with whatever force was necessary to see that no miner was "molested."

The miners' union called for a meeting with the superintendents, but none showed up. As the day drew nearer, Ayers reported rumors of a battle between the union and the counter league, while Forbes reported that the two were negotiating a consolidation. Neither was correct, but quite a number of miners did desert the union.[15]

When the 2d of August finally came and passed without incident, Forbes proclaimed that "the league has been emasculated." Several of the mines resumed work with partial crews at $4 a day; the union made no demonstration; and the trouble appeared to be at an end. Then at 2 o'clock in the morning on August 4th two dozen armed men raided the men working the night shift at the $4 mines. Shots were fired and stones were thrown into the shafts. Several miners who came to the surface were bruised or badly beaten. One, Joseph Gerrans, broke his leg falling into a hole while fleeing the raiders. When he was carried into town later that morning, public indignation was aroused and there was talk of lynching the "cowardly villians."

The miners' union was immediately suspected and Marshal Coleman promptly arrested Leonard and seven other union officers and members. He also seized the union's books and membership rolls. That afternoon the prisoners were turned over to the county sheriff, who took them to Hamilton for their own safety. There, charged with riot, they were crowded into a 7-by-10-foot cell with six other prisoners. Neither Forbes nor Ayers seem to have had any doubts about their guilt. Both accused union members of the raid, although Ayers felt the "night-prowlers" did not represent "the spirit of the Miner's Union."

Forbes took a peculiar satisfaction from the incident, noting "this foolhardy raid upon the workingmen is regarded as a most fortuitous affair in the public interest. The outraged law is now aroused, and popular indignation is boiling. . . . The community is now fully aroused and all seem to be glad that the issue has come so pointedly." In their own defense the accused miners charged that the raid was staged by the owners to discredit the union.[16]

A Citizens' Meeting was called the following evening primarily to assure outside investors that order had been restored in Treasure City. An executive committee of twenty-five was also appointed to aid the city and county officers in maintaining the peace and a number of watchmen were put on guard against possible incendiaries.[17]

A few days later, on August 6th, the miners' union met at Smith's Theater, disavowing any knowledge of the "wanton assaults," and calling on the public and especially the press for a "suspension of opinion until a hearing is had for our brethren, and all testimony elicited, showing their guilt or innocence." Then the chairman, John Shannon announced that despite the many benevolent activities of the union, "a few evil words spoken on the outside had placed them in a false position before the community, and censure and public distrust was the result." Thus he suggested disbanding for a time to wait and organize anew. A vote was called and the Miners' Union of Treasure City disbanded.[18]

Still the bitterness remained, as Forbes wrote, "we have been asked to make favorable mention of those who took pains to disband the organization. It is too late . . . our advice had been scouted, and threats had been made; and the bulk of those men hold us no good will now. So far, the case is even—for they owe us none." The citizens' committee was also maintained for a time after the union disbanded to insure that it did not reorganize too soon.[19]

When the accused miners were brought to trial the following day the prosecution had no evidence linking them with the night raid of the 4th. Instead they were charged with rioting on July 27th, when they visited the mines to stop contract work at less than $5. Because of the crowds it attracted, the trial was held on the stage of the New Melodeon Theater. Robert D. Ferguson, former president of the Virginia City Miners' Protective Association, and E. C. Brearly were counsels for the defendants, who were tried as a group. The prosecution presented several witnesses, including Dr. Goodfellow, who identified all but two of the defendants as having taken part in the incident on July 27th. But in cross examination all testified that no force had been used. Late that evening Ferguson summed up the defense, charging that "there was an organized persecution of the prisoners for ulterior purposes, and the shooting into the shafts was done by hirelings, to carry out the designs of those who were behind the curtain." The jury debated for over two hours while the friends of the accused catnapped on the sawdust floor. Shortly after midnight the jury returned, finding all innocent of rioting, but six, including Peter Leonard, guilty of the lesser charge of unlawful assembly. They were each fined $25 plus court costs of $47. Unable to raise the money, they served several days in the county jail.[20]

The miners had started the strike with seemingly broad community support, but in the boom atmosphere of the camp, as soon as the merchants began to feel the loss of the miners' wages, they became impatient for a quick settlement—one way or the other. The raid on the night shift apparently crystallized community opinion against the miners and provided the excuse for legal action to effectively break the strike. This pattern with only slight variation was to be repeated in many subsequent disputes throughout the western mines.

The miners' charge that the night raid was staged by agents provocateurs or by hirelings of the mine owners to discredit the union has some support from circumstantial evidence. If the Eberhardt owners had deliberately precipitated the strike for any of the reasons suggested by the press, then once it had served its purpose they might also have staged the raid to terminate it. Despite the optimism of the press, work had resumed on only a limited scale on August 2d, and without the raid the strike might well have dragged on for some months. The raid

thus brought about the arrest of the union officers, the disbanding of the union, and a clear end to the strike. Finally the state's
failure even to prosecute the union miners for the night raid also
favors this interpretation. If the mine owners did stage the incident to break the union, it was the first use of such tactics in
the western mines, but not the last.

Although many former union miners were seen with bedrolls
on their backs striking out for new camps, the press, at least,
viewed the failure of the strike as the harbinger of good times.
Forbes, having completely reversed his position, now praised
the Eberhardt company for "taking the initiative in the four
dollar movement. . . . It will unquestionably rebound to the good
of the entire community, augmenting our laboring population
at least one-third, and the amount of money expended in the
district by outside parties in a still larger ratio. In fact, we know
of nothing that could have tended more to promote our prosperity, as a people, than this adjustment of miners' wages."[21]

But this wonderful prosperity failed to come. The mines had
been over promoted and their closure during the strike only
hastened the inevitable collapse of the boom. As Ayers later recalled, "there were innumerable mines to sell, but no buyers. . . .
No one could help feeling that the bottom was dropping out of
the district. Business was greatly depressed. Bills came to be
hardly collectible. Everybody almost was living on credit. At
last the crash came. The best houses went into the hands of the
Sheriff, and the town which a few months before was a lively
picture of active prosperity was in the doldrums. Nothing stirred
but an occasional pogonip."[22]

White Pine finally began to revive late in 1870, but by then
miners' wages had been cut to $3.50 a day. The Eberhardt own-
ers succeeded in floating the Eberhardt and Aurora Mining Company, Limited, with British capital to reopen their mines and
Dr. Goodfellow got similar backing for the South Aurora. By
the summer of 1871 more prosperous times had returned to White
Pine. The miners' union was finally reorganized and in May
of 1872 they struck the principal mines demanding a minimum
wage of $4 a day. The local press backed the demand as long

overdue and on June 6th the superintendents agreed to the increase. The union maintained this standard until the virtual abandonment of the camp several years later.[23]

Following the collapse of the White Pine boom, the Austin miners reorganized once again, but with no more lasting success, for the camp was still torn by the frenzy of mining rushes in surrounding districts. On December 19, 1871, they formed the Miners' Union of Austin with Daniel Holland as president. In addition to their benevolent activities, they pledged not to work for less than $4 a day. The editors of the *Reese River Reveille* looked upon the union's purposes as "not only legitimate but praiseworthy" and encouraged miners to join it and the public to support it.[24]

Within three months the union had signed up 130 members. With roughly a third of the miners in the district, they felt strong enough to enforce the minimum wage. Holland argued, "as the Virginia and Gold Hill miners receive four dollars per day, we think that the miners of Austin, coming farther into the interior of the State, where it costs more to get here and at least one-third more to live, should get at least the same." But the bulk of the work in the mines around Austin was being done on contract, so that the companies had few wage employees. In the largest mine, operated by the Manhattan company, there were 112 miners of which 94 were on contract and only 18 were hired by the day. Similarly the Pacific Union Mining Company, Limited, which employed fifty miners, had only ten on day's wages. Still the union was determined to enforce the principle and with so little at stake the Manhattan company's manager agreed to the minimum wage. The British company's superintendent, Henry Prideaux, however, refused.

Thus on February 17, 1872, forty union men, headed by president Holland, marched up to the Pacific company's works to reiterate the demand. The union men felt that Prideaux was very cooperative, offering to hoist up his men and telegraph the directors if the men wanted $4 a day. As the crew came up they were "politely requested" to join the union and all but one signed up.

Prideaux, however, saw things much differently. Claiming

he had been intimidated by the "excited and dangerous crowd," he filed a complaint against the union. The sheriff promptly arrested Holland and twenty-eight other union members. They were charged with unlawful assembly, with stopping the miners from work by "violence and intimidation," and with forcing the miners to their "great terror" to sign a pledge not to work for less than $4 a day. When the case came to trial the following week, charges were dropped against all but Holland and four other union officers. Over fifty prospective jurors were called before the preemptory challenges were exhausted. Even then the jury was unable to reach a verdict, standing nine to three for acquittal, and the case was dismissed. With the failure of the prosecution, Prideaux finally acceded to the demand for $4 a day. But a new mining rush, like a plague, once again decimated the Austin union.[25]

That spring Holland and other Austin miners joined the new rush to Pioche, some two hundred miles to the southeast. Silver had been discovered there several years before, but its remoteness had slowed development until the Raymond and Ely company struck bonanza ore in 1871. Within a year the company's mines became the largest silver producers outside of the Comstock. Their bullion yield reached $12,000 a day and the Nevada State Mineralogist declared them "the most profitable mines in the State." The ensuing rush swelled Pioche's population to several thousand in 1872. Its remoteness also attracted a disproportionate share of gunmen, bunco artists, and gamblers who earned for it the name of "the wickedest place on the Pacific Coast."[26]

Miners had received $4 for eight hours work until late 1872. Then, the day after Christmas Captain H. H. Day, superintendent of Raymond and Ely's "most profitable mines," increased the shifts to ten hours. The men went down grumbling at 7 A.M. and at 3 P.M. they demanded to be brought up, refusing to work more than eight hours. Day discharged them and the night shift refused to go to work. A hundred men either were discharged or struck. That evening a crier went the rounds bell in hand announcing a miners' meeting at the court house. Between six and seven hundred men showed up but only miners were admitted.

There were appeals for support of the strike and talk of organizing a union but no action was taken.[27]

The following morning the Raymond and Ely hired a number
of gunmen from the local toughs "for the protection of the property from any violence which might be offered by those who
had struck and been discharged." With the gunmen on guard
a new crew, willing to work ten-hour shifts, started that evening.
The strikers, not yet organized, were intimidated by the gunmen
and heavy rain further dampened their spirits.

Although the striking miners failed to put in an appearance at
the Raymond and Ely, the gunmen found other targets. Nearly
two weeks before, the superintendents of the Raymond and Ely
and the adjacent Pioche Phoenix had agreed to stop work in a
disputed piece of ground where drifts of the two companies had
come to within a few feet of each other. But on the evening of
the strike, while the Raymond and Ely was abandoned, the
Phoenix men broke into the drift and barricaded it with sacks
of rock. When Day learned of the barricade the next day, he
brought down some of the gunmen and started fortifications of
his own. Shooting began almost immediately. One of the Phoenix
owners, Thomas Ryan, was killed within the first few minutes,
but the fighting continued sporadically until the following afternoon. The gunmen on the top also got their "man for breakfast,"
killing a hapless Norwegian miner on his way to work in the
predawn darkness. These two killings finally stirred Pioche citizens into forming a Citizens' Protective Union to end the "car-
nival of blood," but no one was ever arrested for these murders.[28]

The miners finally organized a union on January 2, 1873, with
Michael Cody president and Daniel Holland, vice-president.
Holland called on all the miners to join the union, explaining its
aims were "merely to maintain the rights of miners, to succor
each other in sickness and to bury the dead. We seek no end
not in full accord with the law and good order. We point with
pride to the Union at Virginia and Gold Hill, and seek to emulate
its example." But most of the footloose miners who had come to
Pioche were much more interested in striking it rich than in
striking for rights. The prospect of having to face Raymond and

Ely's gunmen every time there was a dispute killed the budding union spirit in nearly all the rest. Thus only a fraction of the men joined and by March even they disbanded. Pioche's prosperity was short-lived too. The following year Raymond and Ely's bonanza ore was exhausted and two years later the mine closed down.[29]

The Eureka mines, some fifty miles east of Austin, eventually proved more productive and much more lasting than those of either Pioche or Austin, although their ores were more refractory and lower grade. The rush began there in 1870, but it was not until October 16, 1873, after the collapse of the first boom, that the miners there organized the Ruby Hill Miners' Union. No wage cut was attempted, so the early years of the union were free from any significant labor disputes. With the slow but steady development of the Eureka mines, the Ruby Hill union thus grew peacefully into the largest and strongest Nevada union outside of the Comstock. When Austin finally settled down, a strong union was at last established there in June of 1875. It joined with the Comstock, White Pine, and Ruby Hill unions to provide a firm base for the future growth of the movement in Nevada.[30]

During this same time the mining labor movement had spread north to Silver City, Idaho. These mines, discovered in 1864, were the most productive lode mines in the territory for nearly two decades. An attempt to establish a Miner's League there at the height of the boom in October of 1867 had failed from lack of support in the "every man for himself" fever of the boom. But as the camp became more settled, a union was at last successfully organized in the spring of 1872. The principal mines were the Golden Chariot, Minnesota, and Mahogany, located on War Eagle Mountain, two miles southeast of Silver City. By 1871 they were producing over $1 million in bullion annually and employing nearly four hundred miners, most of whom lived at Fairview, adjacent to the mines.[31]

Wages held at the Comstock rate of $4 a day and were not an issue of dispute. But for some time there had been a growing discontent among the miners with the Mahogany mine fore-

man, John Jewell, whom they accused of "lavish abuse." Finally
they decided to take action, posting crude notices in boardinghouses and saloons on the mountain.

TAKE NOTICE.

War Eagle Mountain miners and laborers: Be it well understood between all parties interested in earning their living in
those mines, we are oppressed with slavery and bad rules since
Mr. Jewell, foreman of the Mahogany mine, has entered this
camp, and we will all unite ourselves together for the sake of our
beloved country and camp to send this nuisance by stage to Win-
nemucca, and never to return no more to this camp under penalty
of death—and all to meet harmed on the divide between Fairview
and Orevena, on Wed. Ev. March 20, at seaven o'clock P.M. To
all nations, kindred and tongues.

Some three hundred miners, massed at the Mahogany that
evening, told Jewell to leave the district by seven o'clock the
following morning "or suffer the consequences." Rumors soon
circulated in Silver City that the miners threatened to hang the
foreman and burn the works. That night acting superintendent
George Coe sent men up War Eagle Mountain armed with shotguns and rifles to protect the mine. For good measure he also
sent along two mountain howitzers—one twelve- and one six-
pounder—with solid shot, grape, and canister.

The following morning, Jewell, safe behind the barricades
of "Fort Mahogany," announced that he would not leave. The
miners were equally determined, however, and they resolved
not to go to work in any of the mines as long as Jewell remained.
This action proved effective. The other superintendents pres-
sured Coe; the next day he fired Jewell; and the strike ended.
Jewell argued that the trouble was trumped up by a few outsiders, but the local editor concluded "when all the miners in
camp, with scarcely a dissenting voice, express a universal dislike to a man, there must be something wrong."

On March 21st, even before the strike had ended, the miners
met again to organize as the Fairview Miners' Union. By the
following day 250 men had signed up. The constitution and
bylaws, published in full in the local paper, were identical with
those of the Gold Hill union. The editor supported the union,

although he cautioned them not to "meddle with matters, which may concern them, but with which they have no legitimate right to interfere." The miners, however, had found new strength in union and they continued to take strong action whenever they deemed it necessary. The Fairview Miners' Union thus became one of the most active unions outside of the Comstock, and it prospered for several years before the failure of the mines forced its decline.[32]

The miners' union movement in the meantime had also begun to stir in the old gold camps across the Sierra in California. There in more settled communities it met with more solid support from both the miners and the community, whose fortunes depended on those of the miners. But there too the opposition of the mine owners was much stronger; the issues were more complex; and the underlying antagonism between the workers and the management was much more intense. With malevolent chemistry these combined to make the two major disputes there the longest and bitterest of any during these turbulent years. But in the end the miners emerged victorious.

Although placer mining still produced most of California's gold, profitable lode mining had developed at several points, most notably at Grass Valley on the western slope of the Sierra and at Sutter Creek in the foothills along the Mother Lode. Miner's wages here had declined to $3 or less a day, much lower than in the silver camps, and lower living costs only partly offset the deficit. The stability of these more settled camps was probably their greatest attraction to most miners with families. To some, however, the attraction was the free gold that laced the quartz lodes and offered much higher "wages" to those who wished to engage in high-grading or, as it was less euphemistically termed, ore stealing.

If a miner chose to tuck away just a fraction of an ounce of free gold in his pocket every now and then, he could raise his daily wage to any rate that his conscience, or the careless eye of his foreman would permit. Indeed it took a strong man to resist the temptation. As one mining editor observed,

miners are not more dishonest than men of other avocations; but the
temptation to annex a small piece of ore rich in precious metals seems
too much for the average underground man, when, by his own hand,
he exposes such treasure down in the dark galleries of the mine. He
has been taught to believe that it is the duty of the mine officers, from
manager to shift boss, to look out for such things. If he thinks he will
be able to appropriate a few pieces without detection, he usually
yields to the temptation.[33]

The law also provided some shelter for the tempted, since to
break a piece of gold-bearing quartz from a mine and carry it
away without the owner's permission was not legally considered
theft, just trespass. Only after the rock had been broken down
by the owner was its removal considered a theft. It was usually
impossible, however, to get evidence that would satisfy a jury
of the latter crime.[34]

Even if the miner did resist temptation, he was still suspect.
For when he found rich ore and turned it in to the office, the
natural question was, how much had he held out? Thus many
concluded "they may as well have the game as the name." Their
attitude was clearly expressed in a conversation that developed
when a superintendent asked a Cornish miner just going on
night shift,

"If you find a pocket here tonight, how much will you turn over to
the company?"
"That," responded the miner, "depends on how big she is. If we
can't get away with 'er, we'll give'n all up, but if she is a little one
us'll keep en all."
"That would not be fair," said the superintendent, "you should give
the company at least half."
"No," said cousin Jack. "Suppose us'll give you a lot of gold in the
morning. What do you say—where's the rest of un—that's what you'll
say. No, us'll give'n all or keep en all."[35]

Although there is considerable question as to how prevalent
high-grading actually was, there is no question that it was a
major source of friction in the California lode mines. The mine
owners claimed that it took a heavy toll on their profits and
charged that miners' unions and strikes were aimed at protect-
ing it, while the miners felt that the procedures that the owners
introduced to curb it were offensive and degrading. The result-

ing distrust on the part of the owners and resentment on the part of the miners were the seeds of much of the labor difficulties in the lode mines.

The first miners' union in California grew out of a dispute ostensibly over the introduction of dynamite at Grass Valley, but high-grading lurked in the background. Dynamite was first introduced on the Pacific Coast under the name of Giant Powder in 1867, just a few years after its discovery by Alfred Nobel, and it was used with great success by the Chinese miners employed in tunnel work on the Central Pacific railroad. The mining superintendents in Grass Valley district began experimenting with it in the latter part of 1868. Although at $1.25 a pound giant powder was much more expensive than black powder at 11¢ a pound, some mining superintendents found that it could reduce the cost of ore extraction by changing the system of mining. Where two men were needed to drill the large holes required for a charge of black powder, a single man could drill the smaller holes that sufficed for the more powerful giant powder. This change from double- to single-handed drilling reduced the mining crews by nearly half, more than offsetting the higher cost of the powder. The managers also claimed that its use put an "instant end" to high-grading, since with giant powder the employment of just one or two men in each shift for the special duty of firing the blasts and cleaning up the stopes prevented all the rest from having access to the newly broken rock.[36]

The miners however were opposed to the introduction of giant powder and of single-handed drilling for a variety of reasons. It is difficult to say which they considered the most important. They complained that in the poorly ventilated mines the "noxious gases" produced by giant powder caused headaches, nausea, or even "something worse, which baffles the skill of the medical fraternity." Subsequent medical study confirmed its ill effects. The miners also complained that the change from double- to single-handed drilling both reduced the number of miners and threw the business open to unskilled laborers—especially Chinese—since single jacking required much less skill than double jacking. Moreover, as far as the Cornish miners were concerned, the question of the relative efficiency of the

two techniques had been settled in favor of double jacking long
before in the tin mines of Cornwall, and they resented its being
raised again. How effective the change to single-handed drilling
was in curbing high-grading is hard to assess.[37]

The dispute erupted on April 21, 1869, when William Tisdale,
superintendent of the Star Spangled Banner mine, decided to
change over exclusively to giant powder and to single-handed
drilling. The full crew of twenty-five miners refused to work and
Tisdale discharged them. He promptly hired a new crew but a
few days later he discharged them for the same reason. Within
two weeks two other superintendents, S. W. Lee of the Empire
and J. H. Crossman of the North Star, also switched to giant
powder and their combined crews, totaling over two hundred,
quit work. The other superintendents, however, saw no advantage in giant powder and continued with black.[38]

Led by Thomas Faull, the striking miners concluded to form
a "branch league" of the Comstock unions and they telegraphed
the Gold Hill union for a representative to help them organize.
On May 10th they issued a resolution that not only reaffirmed
their opposition to the use of giant powder and single-handed
drilling, but further pledged not to "allow" anyone to work
underground for less than $3 a day. Although underground min-
ers were already being paid this rate, muckers and carmen were
paid only $2.50 to $2.75.[39]

The demand for a minimum wage greatly complicated the
dispute. Amos Morse, editor of the *National*, promptly cautioned,
" 'Allow' is a big word . . . forcible resistance will be foolish and
fruitless; owners and superintendents may be persuaded but
they will never be intimidated. We advise you to keep cool, let
things take their course, and attempt no strikes." He urged that
they take no action until the merits and hazards of giant powder
had been properly evaluated.[40]

A hardrock Hamlet offered similar counsel in his soliloquy.

> Black powder, or giant powder? that's the question.
> Whether 'tis nobler for miner to suffer
> The Superintendents to carry on their mines,
> Or throw down their picks and hammers,
> And by "striking," dictate to them. To strike, to league:

To work no more; and by a "union" end
The backache and the thousand ills
Miners are heir to; 'tis a consummation
Devoutly to be wished. To strike,—to league,—
To league, perchance to fail; ay, there's the rub,
For if we fail, what evils then may come,
When we have shuffled off our daily wages,
Must give us pause; there's the coin
With which we buy our daily grub;
For who would quit and willfully discard
The steady work, the payday coming sure,
The trust at grocery stores and butcher-shops,
And see his family suffering for bread,
When he himself might in prosperity live,
By using giant powder? Who would these fardels bear,
To grunt and sweat under a weary life;
But that the dread of "getting broke,"
That well-known state from whose bourn
Our wages keep us, puzzles the will,
And should make us bear those ills we have,
Than to fly to others we know not of?
Thus common sense doth make cowards of us all,
And thus the native hue of miners' leagues
Is sicklied o'er with the pale cast of thought,
And strike o' little pith or moment,
With this regard their currents turn away,
And lose the name of action.[41]

On May 14th, James Brew, vice-president of the Gold Hill
Miners' Union, arrived in Grass Valley to officially organize the
Miners' Union of Grass Valley District. Phillip H. Paynter was
elected president and George Taylor, formerly of the Gold Hill
union, was vice-president. Faull was not elected to any office,
and, apparently at Brew's suggestion, the resolution of May 10th
was repudiated to the extent that the union would not oppose
single-handed drilling, nor forcibly interfere with the management of the mines. The strike was thus limited to two issues;
the miners' refusal to use giant powder because of "its alleged
injury to health" and their demand for "$3 per day for all men,
no matter how employed, who work underground."[42]

Most of the companies agreed to both demands, but the
minimum wage issue did cause two others to join with the three
giant powder mines in opposition to the union. Anticipating no

further "interference" from the strikers, Lee of the Empire advertised for a hundred miners and announced that he would resume work on Monday May 24th under the old conditions.[43]

Early that Monday morning nearly a hundred miners, headed by George Taylor, assembled at the Empire hoisting works to warn men intending to start work not to do so. The striking miners indulged pretty freely in threats and many carried pistols. About forty men had signed up for work, but only a dozen went to work. They were warned to collect their pay at the end of the day.[44]

A citizens' meeting was called that evening. After hearing Lee's version of the incident a committee, headed by State Senator E. W. Roberts, was appointed to draft resolutions to be voted on the following evening. George Taylor and a number of miners attended the second meeting. They presented the union's position, and helped defeat a resolution charging the union with unlawful assembly and threats of violence. Other resolutions were then passed, supporting the "sacred and inviolable" rights of the mine owners to manage their own property and the miner to make his own contract for his labor. Before the meeting adjourned a few miners gave personal testimonials of the sickness brought on by giant powder. Senator Roberts suggested legislation to require proper ventilation of the mines, but Taylor replied that "it would cost more to ventilate the mines than it would to work them."[45]

That same day the mine owners lost considerable community support with the publication of a letter from one of the owners to superintendent Crossman. The unidentified owner advised

ultimately the rough work of our mines will be done by Chinese. For the present the policy of "masterly inactivity" will receive the indorsement of owners here, and should be adopted by all of your mining superintendents. Two months will suffice to weed out a class of miners who, in your district, have for the past fifteen years, through specimen stealing and high wages, arrived at the conclusion that they have a right to dictate to owners how their property should be managed. A new element must be introduced.

Morse of the *National* denounced this as "a contemptible and slanderous libel" on the miners. He also noted that the introduction of Chinese miners at reduced rates would cost the local merchants more than $10,000 a month in business and reminded them that

Grass Valley lives not off the mines, but the miners—an important distinction. The wealth of the mines all goes to San Francisco. The wages of the miners are spent here. . . . Mine-owners have the un-questionable legal right, if they choose, to employ Chinese labor. It may possibly be for their own interest, which is all that any of them, except the few who own property and live in this county, care a copper for. As long as they get their dividends, they care not whether a church, school-house, hotel, store, library, or anything else exists in Grass Valley or Nevada.[46]

Despite such criticism, Crossman maintained that "masterly inactivity" was the only course of action against the "loud mouthed agitators and troublesome fellows, who by dint of bravado and cunning, have succeeded in leading the masses."[47]

The strike in fact dragged on for more than two months. Lee continued to employ whatever giant powder men he could, but he never got more than eighteen at a time although he adver-tised for as many as 135. The only violence in the strike occurred when two of Lee's men were waylaid shortly before midnight on July 10th. Four club-wielding assailants jumped them on a nar-row trail near the Empire mill and beat up one while the other escaped. Lee offered a $500 reward for the conviction of the assailants and promoted a short-lived Law and Order Associ-ation to investigate.[48]

Morse of the *National* condemned the assault but assured his readers that the miners' union was in no way involved; "its mem-bers are as peaceable and good citizens as the members of any church in the State." The editor of the *Alta California*, how-ever, not only accused the union directly but leapfrogged the charges, finding accomplices even among the owners of the Eureka and other mines which "continue to employ members of the Miners' Union, and give them money which goes to pay for the outrages." Morse denounced this "Altaic" logic as a concatenation of "the doltish stupidity of the ass, the blunder-ing ignorance of the beetle, and the silly loquacity of the par-rot." He then gave his full support to the strike against giant powder, arguing "the miner's health is his capital and he has as much right to jealously guard it, as the rich man has to protect his wealth."[49]

The solidarity of the miners in refusing to work with giant powder, the lack of solidarity among the owners in demanding its use, and the general community support for the miners finally brought victory to the union. On July 13 the Star Spangled Banner resumed work with black powder. A week later the Empire switched from giant powder to gun powder. The North Star remained closed until August when Crossman was replaced by a new superintendent who reopened the mine with black powder. The problem of high-grading was attacked directly by the introduction of changing rooms in which the miners stripped out of their work clothes, walked naked across the room, and put on their street clothes. Although this did not end high-grading, it substantially reduced the losses.[50]

The Grass Valley union prospered and its membership grew to more than seven hundred. Following the strike its energies were directed to the disbursement of sick benefits, the acquisition of a hall, and the election of its president as justice of the peace. Its members briefly indulged their aggressions in anti-Chinese agitation.[51]

There were no further attempts to introduce giant powder in Grass Valley for the next three years. But during that time giant powder was introduced without opposition on the Comstock and elsewhere in the West. Its use proved to be more economical to the mine owners and no serious health hazard to the miners when adequately ventilated. The noxious gases, which caused "dull headache, drowsiness and sundry nervous complaints," were found to be a hazard only when burning was incomplete and the miners reentered the area before the air had cleared. At the same time at Grass Valley work in the Eureka mine moved into barren ground; dividends were suspended and assessments levied.[52]

Thus in January 1872 the Eureka superintendent, William Watt, who had led the companies opposed to giant powder, concluded to give it a second try. The miners' union met on February 3d and although many of the miners still opposed giant powder they could not agree on whether to strike the Eureka. The question of a strike was deferred to the next weekly meeting and then to the next. When the meeting of February 17th still failed to bring in a vote for a strike, a few miners decided to take the matter into their own hands. Shortly after eleven that evening two giant powder men on the night shift in the Eureka were ambushed as they returned to their boarding-house after work. One of the men was hit in the chin, hand, and leg by shotgun blasts; the other escaped injury.[53]

The editors of the Grass Valley *Republican* charged that the attack was part of a "dark and atrocious" conspiracy like the Ku-Klux Klan. Others joined in the chorus, calling for a vigilance committee to "drive the wretches from our midst, or hang them high as Haman." The union's prompt denial of any complicity in the attack, however, was generally accepted by most of the community. The two ambushed miners left Grass Valley a few days later but the rest of the Eureka giant powder crew were not intimidated.[54]

In the meantime the owners of the Empire also decided to use giant powder again. But when the miners struck in protest, their new superintendent, Jack White, organizer of the Comstock unions, telegraphed the owners to reconsider, at least temporarily. The following day the miners were back at work with black powder. The *Republican* considered the Empire's retreat to be "in singular bad taste," complaining that such a "cringing, fawning, yielding policy in the giant powder matter for years past has paralyzed the mining interests of Grass Valley and prostrated all branches of industry."[55]

The Empire's retreat also prompted some of the miners to finally strike the Eureka on March 1st. But it was a last ditch effort and the majority of the union men, refusing to support the strike, continued work. Their refusal reflected the growing realization among most of the union miners that giant powder was, in fact, not permanently harmful and that its introduction was inevitable. The latter conclusion apparently was shared even by many of those, still opposed to giant powder, who began leaving Grass Valley for the Comstock and other Nevada silver camps where it was already in use.[56]

Jack White, in fact, citing the extensive use of giant powder elsewhere, finally convinced the miners at the Empire to change their minds and accept it too. Thus the opposition to giant powder collapsed and on March 15th the miners' union rescinded their resolution against it, leaving members "free to act as they please in the matter." The union's acceptance of the evidence that it was not a serious health hazard and their decision to oppose it no longer, further strengthened the union's position in the community, winning them praise even from such critics as the *Republican*. With this issue finally resolved the Grass Valley miners turned their attention to benevolent activities and the maintenance of a minimum wage.[57]

The expanding mining labor movement met its strongest opposition on the Mother Lode in the "Amador War" of 1871. Here again the issues were complex and high-grading again lurked in the background, but here too the miners' solidarity and their solid support within the community carried the day. Here too for the first time in the West, the mine owners called in the state militia in an attempt to break the strike. Although this intervention proved more comic than tragic, it set a precedent for corporate manipulation of state and federal troops in their fight against the mining labor movement for decades to come.

Amador County boasted the three most productive mines on the Mother Lode: the Amador and the Oneida, just south of Sutter Creek, and the Keystone, two miles to the north at Amador City. In 1870 these mines employed some three hundred miners and surface workers and produced over $3,000 a day in gold—a third of which was paid out in dividends. The most profitable were the Amador Mining Company's mines which had made their original owner, Alvinza Hayward, one of the richest men on the Pacific Coast. They were purchased by the company in October of 1867 for $563,000 and within two years had paid back more than that in dividends.[58]

Despite the prosperity of the mines, miner's wages in Amador had slowly deteriorated. The miners, however, did not attempt to organize until the Amador company's new superintendent, Captain Hall, cut surface wages to $2 a day in the summer of 1870. Hall showed open contempt for the miners and took a peculiar delight in humiliating them as much as possible, asserting that tea, coffee, meats, puddings, and so on were "luxuries much too good for their class." This attitude also offended many in the community.[59]

On July 2, 1870, the miners of Sutter Creek and those sympathetic with them formed the Amador County Laborers' Association. As set forth in its constitution, "the objects of this Association are for the protection of white labor, to maintain its dignity, to secure a fair compensation therefor, and to discourage competition of inferior races." Within a month nearly four hundred men had joined. The president, Luke W. Byrne, was a former member of the Virginia City Miners' Union, and the secretary, John A. Eagon, was a vigorous, but humorless, Virginian who had come to the county as a miner in 1851 and later turned to law and politics. Apparently as a result of pressure from the Laborers' Association, Hall resigned the superintendency in August.[60]

The local press supported the association, but the editor of the *Amador Ledger* did caution it against becoming "a political machine to aid men to office whose hands were never calloused by honest toil." But when the mine owners entered local politics in a scheme to cut their taxes, the association also entered the political ring. The Amador Mining Company sought the privilege of determining the assessed value of its own property. The county assessor had valued it at the purchase price of $563,000 and the company wanted to slash this to a scant $175,000. They put up mining superintendent R. C. Downs, as a candidate for county supervisor, pledged to secure the reduction. Backed by the association, John Eagon defeated Downs and the tax scheme, saving the county some $20,000 in taxes.[61]

Despite Captain Hall's resignation, the miners' lot did not improve much. The new superintendent, John A. Steinberger, made no further wage cuts but he cracked down on highgrading. A changing room was set up where the miners were compelled to strip down on coming out of the shaft, step across the room, and put on other clothes, leaving their work clothes to be examined by the inspector. Many found this procedure offensive, arguing "that there was no law in the United States compelling a man to expose his person." But a large number approved of the system. On the Comstock the miners' union had argued for changing rooms to allow the miners working in the hot, steamy depths to change into dry clothes before facing the icy blasts of the Washoe Zephyr. The association thus made no protest against the changing room and the mine owners argued that only the guilty objected.[62]

With curtailment of high-grading as a relief against low wages, however, the miners now took a vital interest in establishing a minimum wage. Thus in May of 1871 the association announced that wages in the mines and mills were insufficient to support a single man, much less a family, and they demanded, "For first hands and engineers not less than $3 per day, and for second hands not less than $2.50 per day, ten hours to constitute a day's work." A committee was appointed to notify the superintendents.[63]

Two of the superintendents readily agreed. But all the others positively refused, even though most had only a few bucket men working underground for less than $2.50 a day. The schedule demanded by the miners made very little advance over the existing rates, but the managers claimed "the right to make even a small advance implied a right to control the working of the mines." The owners still smarting over the association's successful opposition to their reassessment scheme, were in fact opposed even to recognition of the association.[64] The miners thus

RESOLVED, That on the 1st day of June the members of this association go in a body to the mills and mines in this county, that refuse to pay the wages established by this association, and notify all persons who may be at work to stop working until the superintendents agree to pay such wages, and that the investigating committee be and is hereby authorized to draw on the treasury of the association in favor of any member who is not able to support himself while out of employment; and that no member of this association will work in such mines or mills until such wages are paid.[65]

Two to three hundred miners, led by Byrne, Eagon, and county treasurer James Meehan, visited the mines around Sutter Creek, ordering all the men to quit work in those paying less than the minimum wage. They closed the Amador, Oneida, Keystone, and half a dozen others, but allowed the engineers to remain on the pumps to keep the mines from flooding.

The preceding evening General David Colton, president of the Amador company, had arrived from San Francisco to confront the miners. In his youth as sheriff of Siskiyou County, Colton had outgunned an angry mob of miners intent on freeing a man from the county jail. For his bravery Governor Bigler had dubbed him an honorary brigadier general in the state militia—a title he still clung to. But his past glories carried no weight with the miners at the Amador mine. When he refused to pay the minimum wage, the miners pushed him aside and hoisted the men out of the mine as he stood by flushed with rage and "forbade their trespassing." He felt that he had achieved one victory, however, in having locked up the bullion in his safe before the miners came.[66]

The local press supported the miners' bid for better wages, but lamented the strike, arguing, "the difference between the miners and the owners of the mines is but trifling, and should have been arranged without a resort to such measures. . . . Times are hard, money very scarce, and the withdrawal of the amount from circulation that is paid out by these mines for labor weekly will produce a state of things that this community is not prepared to meet at the present time."[67]

Colton and the other owners were determined to make no concessions and it was rumored that they would not rehire any member of the association. When they did not attempt to resume work, it became apparent that they simply intended to outwait the miners. The miners, however, decided to try to force the companies into compliance by stopping the pumps and allowing the mines to flood. Thus on June 17th a committee of miners ordered the engineers to quit the pumps and the mines slowly began to fill with water. This soon brought action, but not what the miners expected.[68]

On June 21st the California State Democratic Convention, attended by Eagon and Meehan, gave Governor Henry H. Haight the nomination for a second term. That same day General Colton, together with James M. McDonald of the Keystone and James Morgan of the Oneida, asked the governor to call out the

state militia. Although the owners had pleaded to the tax assessor that their mines were worth only a few hundred thousand
dollars, they now swore to the governor that they were worth
"not less than several million dollars." They also swore that "a
riotous body of men, has taken actual and forcible possession
of said properties, contrary to law, and prevented and now prevent the working thereof in any manner; has driven off laborers,
miners and engineers, and other employees therefrom, and refused and now refuse to permit the owners of said properties to
work or use the same in any manner whatever, and now threaten
absolutely to destroy the same." [69]

Since the militia lacked breech-loading rifles and its funds
were exhausted, the mine owners even promised to assume the
costs of arming and paying the troops, claiming that the accumulation of water in a single tunnel would cost them far more.
The governor hastily agreed, on the condition that they pledge
to pay the entire cost. General Colton and his colleagues signed
the pledge and the governor instructed his Adjutant General
Thomas N. Cazneau to call up two infantry companies of the
militia "to preserve order and maintain the laws" in Amador
County. Cazneau, "a spare man with hatchet features and a lim-
ber tongue," had made a small fortune as a marine insurance
adjuster, selling out to the highest bidder, and he had won his
militia post as payment for his services as chairman of the State
Democratic Committee. [70]

Although there were militia companies in the Mother Lode
camps and the adjacent valley towns, Cazneau picked companies from San Francisco, nearly two hundred miles away,
feeling that the closer units might be too sympathetic to the
miners. The call-up of the militia caused great excitement in
San Francisco. Rumors spread that a bloody collision had occurred at Sutter Creek. But word that the mining companies
would pay the troops $5 a day filled the armories with "self-
sacrificing patriots" ready to stand in for any militiaman unable
to go. A battalion made up of the National and Sumner Guards
was formed at noon on the 22d. The mining companies provided
new breech-loading Henry rifles, capable of firing sixteen rounds

a minute, and six thousand rounds of ammunition. "The detach-
ment," enthused one reporter, "is drawn from the first ranks of
social life—young fresh, ardent and ambitious." Nearly a third
of the 177 officers and men were clerks and bookkeepers. The
commander Major John F. Bronson was the assistant secretary
of the State Board of Harbor Commissioners and his Adjutant
Aquila W. Hanna was a law student.[71]

The troops, munitions, and provisions, including "several
queer-looking packages marked 'Coates' Wine Rooms,' 'Whit-
ing's Medicinals,' etc.," were loaded aboard the river steamer
Yosemite. War correspondents of the San Francisco dailies and
the New York *Herald* joined the expedition. The editor of the
Alta California cheered them on, pleased that "the communists
of Amador will be taught to feel that they cannot arouse the
patriotism and love of liberty of the people by acts of lawlessness and violence with impunity." But the editors of the *Chronicle* took a less inspiring view; "the call for military aid is to
sustain the mine-owners in putting men at work who will accept
the pittance offered for one of the most exhausting and dangerous of occupations."[72]

The call-up took the citizens of Amador County by surprise.
As Governor Haight later stated, "It was not our intention to
give any publicity to the matter originally, calculating that the
troops would reach Amador before the fact of their proposed
departure was learned." But the news did reach the mines well
before the troops and George Durham, the county sheriff, rushed
to Sacramento to protest. He told the governor that the mine
owners had not called upon him, that they had "very much
exaggerated the affair," and "that the civil authorities would
be sufficient to prevent any serious trouble." The governor thus
finally decided to halt the militia at Sacramento and go to
Amador to hear both sides. On learning that he had not already
investigated, the *Chronicle* asked, "Is capital so potent that it
requires but to make the demand, and, without any inquiry or
investigation, the Governor orders out the military to suppress
a riot or crush an insurrection which never had an existence
except in the groundless fears of mine-owners?"[73]

Similar sentiments were muttered by the troops when they learned the news on arriving at Sacramento. They made camp at the race track, naming it Camp Latham in honor of one of the Amador mine owners. Here they whiled away two days, alternating drills and footraces with refreshments at the track saloon. Adjutant General Cazneau joined the troops at the track in a jumbo tent protected by his personal six-pound cannon. He also added his son, a military school cadet, to the payroll as aide-de-camp with the rank of captain. The Sacramento *Reporter* grumbled that the whole farcical business "was only done to give a few peacock feathers a chance to flaunt at public expense." The local militiamen also criticized them for going "to fight for the Bank of California." One confided to the *Alta's* war correspondent, "I'm one of the officers of a company numbering sixty-three men, and not ten men of the whole number can be hired to go." "Why not?" asked the correspondent. "Because they don't believe in putting down a lot of poor, broken-down miners, who haven't got a dollar to their name, and are standing up for a living." [74]

The Governor in company with mine owners, Milton Latham and James McDonald, arrived in Sutter Creek on Saturday afternoon, the 24th. He found it warm, dusty and peaceful. It was soon "noised around that the man who stopped the troops had arrived, and that he was bent upon having peace if it took all summer with 'Sheridan ten miles away.'" The miners welcomed the Governor as an arbitrator, cheering and applauding as he arrived at a special meeting of the Laborers' Association that evening. He told them that he had "come among them at some personal inconvenience to himself . . . to see what could be done toward effecting a reconciliation." But when he advised them simply to go back to work at the old wages, they shouted, "No, we will hold out!" The Governor grabbed his hat and left in a huff. The miners then went into secret session, vowing again to "hold out until the owners make a proposition." [75]

The Governor sent a terse telegram to General Cazneau, "Come up with your companies immediately"; then he left Sutter Creek. On Sunday morning the troops, baggage, wagons, and teams were loaded on a special train of the Sacramento Valley Railroad, which hauled them some forty miles up into the foothills to Latrobe. There they began a twenty-mile march to Sutter Creek early in the afternoon. The mine owners again tried to keep the troop movement a secret, buying up all newspapers that came into Sutter Creek. But the word got around nonetheless and John Eagon rode out to see the advancing army.

Since none of the wagons had springs, most of the soldiers chose to walk, and everyone and everything was soon caked with dust. To quiet some of the grumbling, the General treated the men to a cup of wine at Oakdale, but marched on into the night without pausing for food. When the lights of Sutter Creek came into view, General Cazneau ordered his six-pounder and all of the rifles loaded and sent skirmishers ahead to reconnoiter. They met only a curious crowd of onlookers who had waited up for their arrival. The General deemed it safe to proceed, and just before midnight the Amador Battalion captured Sutter Creek "in fine style, with drums beating and colors flying."[76]

General Cazneau immediately went into conference with General Colton, and John Eagon paid them a brief visit to assure them that the miners "would not devour the troops." The miners showed more contempt than hostility for the counterjumpers turned soldiers. One boasted that he "chawed just such men every morning for breakfast," but declined the opportunity of chawing on one right then. Instead the troops chawed down on dinner at the hotels, then bedded down at the skating rink.[77]

The next morning, General Cazneau set up his headquarters in Colton's house while his troops made camp on the Amador company's ground nearby. This post, christened Camp Colton, was in a commanding position above the town. Two outposts camps Morgan and Wright, were also established at the Oneida and Keystone mines. With the troops on guard the engineers started up the pumps again. The water in the Amador was only three feet deep in the lowest level and superintendent Steinberger assured a correspondent that "no great damage has been done."[78]

The presence of the troops only strengthened the determination of both sides. The mine owners, meeting at Colton's house, vowed that "the men who have struck shall never be employed in the mines again." Moreover, they told the *Alta*'s war correspondent that "the matter had reached such a point as to be of vital importance to every citizen of the State, and as such should be resisted, regardless of the risk to individuals and property. . . . If necessary, the soldiers will be kept here for six months or longer." The miners on the other hand had been offered financial aid from the Comstock miners' unions and were likewise determined to hold out to their "last biscuit." They felt that time was on their side, arguing that since "the soldiers do not live at Sutter Creek and they do, they will just patiently bide their time and let anything happen that may, without any interference, until the soldiers are removed; and, when that time comes, they will take the same steps again." Thus one war correspondent concluded with some disappointment, "there apparently is not the most remote idea of any collision with the military on the part of the miners."[79]

With the troops on the ground the owners expected to resume full-scale operations. But although they advertised for men all along the Mother Lode and hired every greenhorn that came along, they were unable to get more than a skeleton crew. Even those who did come to work soon quit. Some were intimidated by the striking miners while others, unaccustomed to mining, simply found the work too hard. The owners published threatening notes allegedly sent to their workmen. One, signed "A Bitter Enemy," menaced, "You ——— ———, if you go to work you will be missing one of these days, and your soul where it should be. So look out and suffer the consequences, for so sure as you do, so sure will you get a bullet through your dirty carkis." The union disavowed such threats, and some must have wondered that a man who could spell "consequences" couldn't spell "carcass."[80]

For the first few days General Cazneau, still fearful of an attack by the miners, kept his six-pounder loaded and ordered his men to build barricades and sleep on their guns at night. Two suspicious-looking miners were held overnight at camps Morgan and Wright but no attack came. The General's son, decked out in a major's uniform, heroicly ordered a sergeant to confiscate a rabbit hunter's shotgun as a "trophy of war." But he panicked under fire. When someone set off a few torpedoes in the middle of the night young Cazneau "ran about in a frantic manner with revolver in hand, requesting everyone he met to 'stoph thwat firing.'" He was later locked up in the guardhouse when he insisted on entering camp without the countersign. Poison oak and diarrhea proved a worse threat to the camp than did the miners.[81]

Eventually the troops became accustomed to the summer heat in the foothills and settled into a quiet routine of sentry duty, drilling, and latrine digging. But with the dullness of things, they soon began griping about bad grub, favoritism, and the snobbishness of some of the officers. Finally on July 3d the troops staged a strike of their own, marching to the General's house and demanding "more bread and no more pork!" They were told to keep quiet and they could have "anything under heaven."[82]

Since the miners were determined simply to outwait them, the troops were of no real benefit to the owners. As Colton soon found, even with them on guard, he was still unable to recruit a large enough crew to resume full operation. Indeed the presence of the troops, which to people outside the district seemingly confirmed the wild rumors of violence, may have hindered the recruitment of scabs, who might have accepted lower wages but not at the risk of their lives. Moreover most of the people in the district resented the presence of the troops and sided more strongly with the miners. Thus through his superintendent, Colton finally opened negotiations with the association. Ironically he ended up conceding much more than he would have if he had settled immediately, since in the course of the strike the miners had increased their demands. After their original demand for a minimum wage for all hands underground, they demanded the same for surface and mill hands, and in response to statements by the owners they further demanded no discrimination against strikers and exclusion of Chinese laborers. On July 5th

the owners finally agreed to all but the minimum wage for surface and mill hands. The Laborers' Association met that evening
and unanimously accepted the settlement.[83]

But before work could resume, James Morgan of the Oneida
withdrew from the agreement, as he wanted to continue employing half a dozen Chinese mill hands. When Chinese labor
became the issue General Cazneau, sensing a potentially disastrous election issue, joined with the miners. He wrote Morgan,
"You will excuse my saying to you, that when it comes to be
understood that the Chinaman is the remaining cause of contention here, you will stand in the unfortunate attitude of opposition to the known and decided sentiment of the good people of our whole State upon that subject." Morgan finally agreed
and the strike ended on July 13th.[84]

The Amador War was over and the battalion broke camp at
Sutter Creek on Sunday afternoon the 16th. A salute from the
General's six-pounder heralded their arrival in San Francisco
two days later aboard the steamer *Capital*. They were treated
to champagne and cigars, and there was even talk of having a
medal made to commemorate the campaign. But the good feeling was destroyed when the troops learned that the mine owners
were withholding $1 a day from their pay for board—that lousy
pork! The State of California was also shortchanged by the mine
owners. Even though the stockholders of the Amador Mining
Company paid out $15,799.25 in "extraordinary expenses," the
owners did not honor their promise to pay all expenses and the
people of California wound up contributing $36,000 to the
adventure.[85]

Although the Amador War itself had been a bloodless affair,
there were three casualties soon afterward. Personal bitterness
between some of the miners and the management flared into
violence on the afternoon of July 24th. Hughey McMenomy, a
member of the executive committee of the Laborers' Association,
and J. W. Bennett, superintendent of the Amador Mill, got into
an argument on the street in front of the Amador company office.
Both drew guns and began firing. The company's bookkeeper,
E. E. Hatch, who had a grudge against McMenomy, rushed out

with his pistol and joined in. The affray ended with both Mc-
Menomy and Hatch mortally wounded. Exactly who shot whom
was not clear. Both wounded men had only fired two shots each,
while Bennett, carrying two pistols for the occasion, got off eight
shots and possibly shot Hatch by mistake when Hatch rushed
into the fight behind him.

Wild excitement prevailed after the shooting. Bennett, Hatch,
and Steinberger barricaded themselves in the company's office
for the night. Frantic telegrams were sent to General Cazneau
to call up the militia again; rumor again spread in San Francisco
that the miners had seized all the mines and the telegraph office; and the newspapers again dispatched their "war correspondents" to the scene. But Sheriff Durham had promptly formed
a posse and closed all the saloons in Sutter Creek. No further
incidents occurred. McMenomy and Hatch died two days later.
Bennett slipped away to San Francisco, and the grand jury later
dismissed charges against him.[86]

The third casualty was Major John F. Bronson, commander
of the Amador Battalion. On returning to San Francisco he fell
ill—General Cazneau surmised, from "the fatigues incident to
camp life and its exposures, the hardships of heavy marching,
and the labors and cares of his position." He died the day after
McMenomy and Hatch.[87]

The Amador War found perhaps even a fourth casualty—
politically at least—in Governor Henry Huntly Haight, whose
misfortune it had been to call up the troops at the ill-advised
bidding of the mine owners in an election year. Although Haight,
a Democrat, favored the eight-hour work-day law, and opposed
Chinese immigration and railroad subsidies, his use of troops in
Amador lost him much labor support. Eagon and other Labor
Association members, active in the Democratic party, bolted the
county convention. Even the Virginia City Miners' Union joined
in, finding Haight "by his *acts* to be an enemy of the laboring
man" and calling for laborers to "retaliate and consign him to
political oblivion." James Phelan, president of the union and a
Republican state senator in Nevada, stumped the California
mines campaigning against Haight.[88]

The Amador War became the subject of several Republican campaign songs. One even credited Haight with planning the whole affair:

Oh, listen to me boys and don't make any noise,
    I am the fighting Governor, you've heard of me before;
I am the little man, oh no! the great "I am,"
    Who squelched the great rebellion way down in Amador.

My star was growing dim, it seemed to me a sin,
    To sink into obscurity and private life once more;
I laid a little plan, like any honest man
    That brought out the great rebellion way down in Amador.

The mill men were my friends, who helped me gain my ends,
    By swearing that the miners would not dig out any ore;
Here was a chance for war, to hear the mob hurrah,
    When I led my gallant army down to Amador.

Then I called out the boys, who liked this fuss and noise,
    To put down those ragged miners who were my friends before.
For if the poor won't work, and for the rich eat dirt,
    I would slay them on their hearthstones, way down in Amador.

When we got to the mines, we found but quiet times,
    The miners never gathered near, except around the store
To buy their week's supplies, and feast their dazzled eyes,
    And wonder why my army came down to Amador.

I had my army there, and then I did not care
    A snap about the miners for I felt rather sore,
For they with eager spite, called me a blatherskite,
    A carpet-bagging scallawag, way down in Amador.

I turned me back for home, my glory all had flown,
    And all I brought to 'Frisco was a sample of the ore,
I was the biggest fool—by folly and misrule
    To lose the chance for Governor way down in Amador.[89]

When the election came in September, Haight, who had won four years before by nearly 10,000 votes, lost by over 5,000—a switch of 15,000 votes. No one issue accounted for his defeat. But his heaviest losses were suffered in the three counties through which his army marched, San Francisco, Sacramento, and Amador where more than 9,000 of his supporters defected.[90]

At the same time the strength of the Amador County Laborers' Association was demonstrated by the election of its secretary, John Eagon, to the state assembly as an independent. The association, having survived its severest trial, faced no further challenge for some time.

Thus California's first miners' unions were firmly established after long and bitter disputes. Doubtless the most significant factors in their success were the internal solidarity of the unions and their strong support within the community. These stemmed from the miners' feelings of permanence in the community and from the full realization by the community that they lived "not off the mines, but the miners." The feelings of community in these older more settled camps contrasted markedly with the "every man for himself" attitudes so prevalent in the more transient boom camps.

Farther east in Colorado, however, the miners' first attempt at organization was much less successful. There too the community was more settled, but hostility toward the Cornish by Irish and American miners and the community at large destroyed the solidarity among the miners and the support of the community. Placer mining had begun in Colorado with the Pike's Peak gold rush of the late 1850s. But deep mining developed slowly, retarded by refractory ores, which required smelting, and by the isolation of the Colorado mines from the growing mining technology of California and Nevada. The mining labor movement in Colorado also suffered from this isolation.

By the early 1870s Colorado mineral production had grown to about $5 million a year, a tenth that of Nevada and about equal to that of Pioche. The most productive mines were on the Gregory and Bobtail lodes in Gilpin County. These mines reached a peak production of $3 million a year in 1871 and then declined to about half that by 1873 as the ore became more refractory. About five hundred men worked in the mines and lived in three adjacent towns: Central City, Black Hawk, and Nevadaville. Wages were only $3 a day but they were high enough to attract newly arrived Cornish and Irish, as well as iron and lead miners from the Midwest. The influx of men from these areas had, in fact, caused a labor surplus that allowed wages to fall. The stiffer competition for jobs here also bred a deep antagonism between the less-skilled Irish and Americans and the more-skilled Cornish, who were most in demand as miners. The local merchants also resented the Cornish because they sent a big part of their wages to their families back home rather than spending it all in the community. This antagonism and resentment undercut any attempt to organize the miners.

The declining yield of the mines, coupled with the nationwide financial panic of 1873, prompted several of the Central City mine managers to seek remedy in a wage cut. On November 10th they reduced wages to $2.70 a day for all men then at work and to $2.50 for any new men hired thereafter. The Cornishmen "threw down the implements of their labor" and met that evening at the Turnverein hall. Joined by other Cousin Jacks working on contract and not directly affected by the cut, they vowed to resist the reduction and set about organizing a union. The Central City Miners' Union was officially formed two days later with Andrew Stevens, president, and William Martin, treasurer. Most of the Cornish miners joined the union, but the bulk of the Irish and Americans, who made up more than half the miners in the camp, refused to join them.[91]

On the eleventh, the union men visited each of the mines, intimidating would-be scabs and forcing a complete suspension of work. They then marched to Black Hawk and Nevadaville to rally support. But even the Cornish there showed no interest in the strike or the union, since their wages were not threatened. Rumors quickly spread that the Cornish would seize the mines and a few of the managers sent frantic calls to the governor for troops. Hired guards were stationed at a number of the mines and one more diabolical superintendent connected a nozzle to his boiler to scald intruders. In the excitement and confusion a trigger-happy guard accidentally shot one of the scabs—the only fatality of the strike. At the prodding of the managers, the sheriff responded by arresting eight union men on omnibus charges of "conspiracy, riot, unlawful assembly, and rout," but they were soon released on bail of $200 each.[92]

Both the miners and the managers refused to compromise, so the battle, which the local editor dubbed "the Cornwall secession movement," settled into a war of attrition. But since the

Central City merchants refused to extend credit to the striking Cousin Jacks, the savings of most were exhausted within a few weeks. The union tried valiantly to start a cooperative store but failed. The managers in the meantime successfully recruited enough Irish and midwestern miners to resume work with the protection of the sheriff. Lacking both community support and solidarity in their own ranks, the miners' cause was doomed. After about a month the union quietly conceded defeat and the Cousin Jacks went back to work at the lower rate. Wages were soon reduced elsewhere in Colorado without further protest. The union continued for a short time after the strike as a benevolent and social club, but it was to be many years before a strong miners' union was organized in Colorado.[93]

If there was any simple pattern to the course of labor disputes during these turbulent years, it was that the mining labor movement generally met with success in the more settled camps but failed in the boom camps. Central City with its Cornish-Irish antagonism was the only exception. These successes in the older camps of California, Nevada, and Idaho apparently stemmed from the stronger feelings of permanence and solidarity among the miners and a broader feeling of mutual dependence and community in the camp itself. Here the miners and the community at large had a long-term stake in maintaining wages and adequate working conditions, and they were willing to fight long and hard for them. The failures, on the other hand, stemmed as clearly from the lack of such feelings in the get-rich-quick, every-man-for-himself atmosphere of the boom camps of White Pine and Pioche, and even the older camp of Austin which was still infected by it. Most of the miners and merchants who followed the rushes to these camps came to make a rich strike or a quick profit, and they had no patience for a long fight over wages or working conditions, if it kept them from quickly earning a grubstake or even briefly curtailed their business. Thus although vigorous efforts were made by some of the miners to establish unions during the booms, they were notably unsuccessful, and it was only after the camps had become more settled and some feeling of community had developed that successful unions were finally formed. This pattern was also evident even in the early mining labor movement on the Comstock with its initial failures in the boom years and its final success once the camp had begun to mature.

This was not a serious limitation on the mining labor movement, however, since it was these more-settled camps that produced the bulk of the mineral wealth and employed the bulk of the miners. Even though the unions still numbered less than a dozen by 1873, they represented nearly half of the hardrock miners in the West and controlled the camps which at that time produced fully three-quarters of the mineral wealth from the lode mines. The Comstock mines alone produced half of the total and half a dozen other camps produced the other quarter. These were also the most industrialized camps, the most highly capitalized, and the most fully controlled by absentee owners. The miners in these camps clearly realized the need for organization to protect their interests; they had the numerical strength and the solidarity to make it effective; and they had the support of the local community which also clearly realized its dependence on their wages. Thus within just six years of the founding of the Comstock miners' unions, the mining labor movement had established a solid base in the western mines and from most of the bitter struggles during these turbulent years it had emerged triumphant.

Union against the Chinese

Within a few years of their founding the hardrock miners' unions were infected with the chronic anti-Chinese fever that had afflicted the Far West for nearly two decades. For a brief time the fever would inflame the unions to violence as the miners panicked in fear that their very livelihood was threatened. But soon it would become only an occasional complaint with the unions, as the miners, no longer feeling personally threatened, came to recognize it as an issue of broader community, and ultimately national, concern.

Chinese placer miners had suffered from persecution by their white brothers since the early 1850s. They had been hit for special foreign miners' taxes by the state, "legislated" out of camps and districts by miners' courts and district laws, forced to pay protection money to local hoods, and driven off their claims and out of their cabins, or even killed, by armed mobs. When they sought jobs in and around the deep mines as miners or laborers, they were usually turned away by bigoted superintendents who felt they could not do the work, or by bigoted miners who refused to work with them. Thus they were generally denied the opportunity to work their way up from unskilled surface workers, to carmen, to muckers, to skilled miners.[1]

Still some did find work in the quartz mines along the Mother Lode in the late 1850s and in the tunnels of the Central Pacific Railroad in the late 1860s. But those superintendents who did hire Chinese miners did so only to exploit them, paying them only $1 to $1.25 a day—less than a third the wage paid white miners. Some Chinese, such as those on the Central Pacific, struck for higher wages, shorter hours, and more "humane" treatment, but they were unsuccessful. Despite such discrimination the Chinese gave a full day's work for their pay and they clearly proved their ability as hardrock miners wherever they were given sufficient opportunity. Their success in the tunnels of the Central Pacific was, in fact, widely publicized.[2]

Thus by 1869 when the transcontinental railroad was nearing completion and large numbers of Chinese were being discharged, mine owners in California and Nevada looked with favor on the prospect of hiring them in their mines and mills. In support of the move they claimed that miners' wages were much too high, and that employment of Chinese miners at $1 a day would allow many mines then lying idle to be reopened at a profit. The miners, however, charged that most of the mines lying idle were so for one of three reasons: they had been worked down to the water level and the owners could not afford to put in pumps; the paying ore had pinched out; or they had failed to pay a profit through ignorance or mismanagement. "Cheap wages," one argued, "would not make machinery cheaper or water less troublesome; would not prevent ledges from pinching out; and would not give brains to those who have them not." Moreover he predicted,

if Chinese labor were to be employed, it would be bad for the general welfare of the town. Where $60,000 are now paid out to white men and mostly expended by them among the merchants and other business men of the town, only $20,000 would be paid out to Chinamen, and not one tenth of it would be spent by them. The $40,000 of balance would, of course, be taken from the people and put into the pockets of the capitalists. . . . Our churches and schools would rot to pieces, our merchants would starve, and instead of a thriving American town, a Hongkong would be established on its ruins.[3]

Although grossly overdrawn, such fears were apparently shared by most of the mining community.

Mining commissioner Rossiter Raymond, however, felt that the cheaper-wage issue was only temporary and would "settle itself by a rise in the demands of Chinamen and a fall in the price of Christians." He predicted that "the Chinese will maintain their hold in this country, if they maintain it at all, not by the cheapness, but by the excellence of their labor." Nonetheless it was the prospect of lowering wages that drew the interest of the mine owners and the opposition of the miners and merchants to the employment of Chinese in the deep mines.[4]

The first attempt to hire Chinese even in the mills east of the Sierra Nevada was thus met with immediate opposition from the miners. The miners' reaction was in fact so swift and decisive, and had such strong community support, that most of the mine owners promptly gave up any further notions of putting Chinese to work in the lode mines. Late in 1868 John C. Fall of the Arizona Silver Mining Company hired about a dozen Chinese laborers in his two stamp mills at Unionville in Humboldt County, Nevada, about a hundred miles northeast of the Comstock. The miners and others constituting about two-thirds of the men in town promptly organized the Workingman's Protective Union. The object of the union was "to protect the interests of the white workingman against the encroachments of capital and Coolie labor; and to use all *legal* means of ridding the country of Chinamen." The leader of the movement was William S. Bonnifield, a former member of the Gold Hill Miners' Union and a brother of McKaskia S. Bonnifield, state senator.[5]

The union demanded that Fall discharge the Chinese and when he refused they boycotted his general store. But this proved ineffective and some of the more zealous miners called for "immediate and forcible expulsion and banishment of all the Chinese now in town." Judge George G. Berry, editor of the town's only newspaper, the *Humboldt Register*, and an Anti-Coolieite, cautioned against running the Chinese out, predicting that it would only bring discredit upon the cause. Moreover Fall vowed that if any union members even came near his mill he would "riddle them with bullets till their hides would not hold cabbages." But the miners paid no attention.[6]

Soon after dawn on the icy Sunday morning of January 10, 1869, sixty armed men, led by a band and banners, marched down Main Street to the houses of the Chinese. There they were met by a number of wagons. Detachments went into each house, ordering the Chinese to bundle themselves and their traps into the wagons in double-quick time. They quietly complied. The wagons then rolled on down to Fall's mills. Fall was away and the Chinese there were also picked up without resistance. In all, forty men and six women were forcibly loaded into the wagons and taken some twenty miles to the Central Pacific railroad at Mill City. A few Chinese were allowed to remain in Unionville just long enough to dispose of property they couldn't take with them. When the mob marched back into town the sheriff notified them that their action was "unlawful" but he made no attempt to arrest anyone. The mob gave three cheers and disbanded.[7]

When the news spread outside Humboldt County, the press denounced the Ku Klux action and demanded justice. A few days later the sheriff finally arrested William Bonnifield and several others. Bonnifield's brother rushed up from the state capitol to defend them and aided by the district attorney, he persuaded the justice of the peace to dismiss the charges on a pledge by the defendants that they would not use force again. The defendants insisted, however, that they were only evicting "a den of Chinese prostitutes" and that they had violated no law.[8]

The federal grand jury meeting in Virginia City disagreed, however. On February 1st they charged William Bonnifield and sixteen others with "the crime of a violation of the treaty between the Empire of China and the United States of America." The Burlingame Treaty, signed by Secretary of State William H. Seward and the envoys of the Emperor of China on July 28, 1868, provided that "Chinese subjects visiting or residing in the United States shall enjoy the same privileges, immunities, or exemptions in respect to travel or residence as may there be enjoyed by the citizens of the most favored nation." On that same date, July 28, 1868, Seward also announced the ratification of the Fourteenth Amendment, which guaranteed to all citizens "the equal protection of the laws." Federal Judge A. W. Baldwin issued a bench warrant on the grand jury's charge and the United States marshal came to Unionville to make the arrests. But Lewis Dunn, the first man to be arrested, petitioned for a writ of habeas corpus, charging that he was "unlawfully" held. McKaskia Bonnifield again came to the defense and Anti-Coolieite George Berry, the district court judge, heard the case. Berry concurred that "no Federal statute has been violated," since "to violate a treaty is not to violate an act of Congress, unless Congress shall have made it a statutory offense and provided a punishment therefor." Thus he concluded the warrant did not charge a crime which could be tried in the courts and it "must be treated as so much waste paper." Dunn then filed suit against Judge Baldwin, his clerk, and the marshal for $15,000 damages for false imprisonment.[9]

Judge Baldwin, outraged at these proceedings, subpoenaed William Bonnifield and others to appear before his court in Virginia City on April 25th. The marshal came to Unionville again, but after serving just a couple of subpoenas he received a telegram from Virginia City to revoke them. Then the United States attorney arrived and J. J. Hill, one of those indicted, was arrested and hustled out of town before he could apply for a writ of habeas corpus from Berry. But when Hill reached Virginia City he applied for a writ from the state supreme court. They rejected his petition after deliberating the case for nearly a month, but at this point Judge Baldwin dismissed the case. As these ill-starred attempts at justice show, the anti-Chinese fever had gained a strong hold even among the judiciary. Moreover the failure to prosecute the miners for the expulsion of the Chinese from Unionville encouraged similar persecution of the Chinese elsewhere.[10]

Attempts at legislative action fared no better. Soon after the incident, Humboldt's assemblyman J. M. Woodworth, prodded by John Fall, had introduced a bill that would clearly outlaw such acts in the future. Entitled An Act for the Protection of Labor, the bill made it a felony for two or more persons to prevent "by force, or show of force, or by threats or intimidations," anyone from being hired, or hiring others, "on such terms as he may choose." But as this could be applied to miners' strikes in general, the Comstock miners' unions gathered 2,700 names on a petition opposing the bill and it was tabled. As a "flank movement" McKaskia Bonnifield introduced a bill to make the employment of Chinese a misdemeanor under the state constitutional provisions prohibiting slavery and involuntary servitude, assuming in effect that all the Chinese in Nevada were "peons or slaves to masters in China." This bill also failed to win the approval of the legislature.[11]

In the meantime the Workingman's Protective Union had bought control of Unionville's only newspaper, the *Humboldt*

Register, adding *Workingman's Advocate* to the title and raising
William Bonnifield to the editor's chair. "It is a long and dangerous leap from the pick to the pen," Bonnifield noted, "but we
have determined to try the experiment, though we shall not pretend to indulge in brilliant rhetorical flourishes or labored dis-
sertations on abstract questions." But as his own words betrayed,
he had already become enamored with rhetorical flourishes and
when his editorial career came to a close in May of the same
year he went into law with his brother. Within six years he
became the district court judge in Humboldt county and his
brother ended up on the state supreme court.[12]

The Union sought a statewide distribution for the paper and
A. C. Hay, president of the Gold Hill Miners' Union, became
the subscription agent on the Comstock. In addition, both the
Gold Hill and Virginia City unions advertised their weekly
meetings in its columns.[13]

The paper's goal was to spread the creed of the Workingman's
Protective Union: "to oppose the employment of Coolie labor,
and by legislative enactment prevent its introduction in our
mines, mills and workshops." But Bonnifield also sensed "among
the working classes, the germ of a great and successful political
party, having for its object the improvement of our condition
and the suppression of Coolie slave labor." Thus he called for
the organization of Protective Unions in every city, town, and
camp in the West, all of which would unite in one grand "Labor
Party."[14]

The Comstock miners' unions enthusiastically joined this
crusade with an impassioned, if baroque, call for a statewide
Workingmen's Convention.

WORKINGMEN OF NEVADA:

To you we appeal in the great crisis now being forced upon us by
the sordid selfishness of the money kings, who, for the lust of lucre,
are striving to check the progress of our young State, and bring ruin
and poverty to ourselves and those dependent upon us for support
and maintenance. Day by day, and little by little, they have advanced
their nefarious scheme. By degrees, and insidiously, has the enemy
drawn nearer; and to-day, from his ambuscade of honeyed seemings,
he rises a monster, which will require our united and firmest action
to defeat. . . .There is danger brooding near; ruin floats on every

breeze; poverty hovers over every rooftree; while desolation awaits anxiously the fruition of the dream of the sordid spoiler—the firm seating of Chinese labor in our midst. . . .Already they have left the railroad lines, and are working in the mills and mines of Humboldt county; and if we defer action, how long will it be before every mill and mine in the State will be worked by Chinamen and *we* be left to starve or seek for employment elsewhere? . . .

The Miners' Unions of this State, well organized and powerful in numbers, are watching with untiring vigils the movements of the foe, prepared to act when the emergency arises. But it is not alone the miner who will suffer in the contest; therefore, we cordially invite all branches of labor to organize and send duly accredited delegates to a WORKINGMEN'S CONVENTION, to be held at VIRGINIA CITY, on the SIXTH DAY OF JULY, A.D. 1869." [15]

Joe Goodman of the *Territorial Enterprise* denounced the call as a "mass of drivel . . . degrading to any man claiming to be an American citizen." Moreover he warned that uniting divergent trades would sow the "seeds of sure and swift destruction" for the Comstock unions. [16]

Despite such criticism the miners and laborers responded enthusiastically. Each union was allotted one delegate for every twenty-five members, The Gold Hill and Virginia City miners' unions sent twenty-one and twenty-two delegates, respectively, the White Pine miners' union ten, the Workingman's Protective Union of Unionville six, headed by William Bonnifield, and the Washoe Typographical Union and Virginia and Truckee Railroad workers one each, making a total of sixty-one delegates representing over fifteen hundred workingmen. [17]

The Convention was held in the district courtroom in Virginia City. Sheriff W. I. Cummings, first president of the Gold Hill Miners' Union, and James Phelan of the Virginia City Miners' Union delivered the opening addresses. Phelan declared that it was the purpose of the Convention "to deliberate on the means of averting the evil which threatened disaster to them and ruin to the State in the shape of the introduction of Chinese labor." But he also hoped "the conduct of the Convention would inspire greater confidence between labor and capital, and elicit the sympathy and approbation of the entire community." [18]

The Convention then went into closed session so the "hard-fisted sons of toil, unused to speech-making and parliamentary etiquette, could express their ideas with more freedom." But when this was all done with, the only action taken by the Convention in this first session was the appointment of two committees, one on permanent organization and the other to draft a constitution, bylaws and resolutions. The delegates then adjourned until the 2d of August.[19]

The second session was attended by additional delegates from the Iron Moulders' Association, No. 185, and the Brewers' Association, both of Virginia City. At this time the Convention officially organized as the Grand Council of the State of Nevada. James Phelan was elected president, and William S. Bonnifield and John Williams, vice-presidents.[20]

They adopted several resolutions echoing the sentiments, if not the verbiage, of the original call. But primarily they vowed "we will exhaust all the resources of petition and argument to check the tide of coolie importation. We discountenance all manner of threats of violence. . . . A braggart's tongue betrays a craven's heart, and ruffianism an unholy cause, *yet, if we are pushed to the wall, then the end must qualify the means*." They left to the Grand Council "all the details of future action." Then with a vote of thanks to the delegate of the Brewers' Association, who provided refreshments for the day, the Workingmen's Convention adjourned till the following year.[21]

In keeping with the spirit of the resolution denouncing "threats of violence" the Workingman's Protective Union had made no demonstration when Fall rehired seven Chinese mill hands. They concluded just to "patiently await the result." They did not wait for long. For Fall soon discharged the Chinese, deciding to economize by increasing the work day to ten hours. The union promptly struck against this change, but they had less support on this issue. After a strike of about three weeks they finally agreed to the longer hours with the understanding that Fall would hire no more Chinese. Thus the threat of "coolieism" was defeated in Unionville.[22]

But within a few months the Comstock miners saw, or at least fancied, a new onslaught on their own bastion. Early in 1869 construction was begun on a railroad, first known as the Virginia and Carson and later as the Virginia and Truckee, to connect the

Comstock mines with the mills on the Carson River. The railroad was controlled by William Sharon and the "bank ring." Giving them a virtual monopoly on the hauling and reduction of Comstock ores, it became a key factor in their fraudulent exploitation of the Comstock mines and stockholders. Like the Central Pacific Railroad, they employed Chinese to grade the roadbed for the track. Starting at Carson the Chinese had worked in the construction gangs for months without incident. But late in September 1869 their work brought them up onto American Flat within sight of the Gold Hill Miners' Union Hall. Rumors then began to spread that once the grading was completed, Sharon would put the Chinese to work in the mines. This was "gall and wormwood" to the miners and the Gold Hill and Virginia City unions promptly met to decide on a course of action. Although in convention they had promised to "exhaust all the resources of petition and argument," they suddenly felt "pushed to the wall" so that "the end must qualify the means."[23]

Early in the afternoon on September 29th nearly four hundred miners, led by their union officers, and a fife and drum, marched through Gold Hill and out along the railroad grade. Before they reached any of the graders they paused briefly to let Sheriff Cummings, a union member, read them an order "to desist from the unlawful business they were on and disperse." The Gold Hill union president, Thomas Atkinson, told Cummings that "they would first do what they had started in to do, and then disperse." The deputy sheriff then read the riot act. The miners responded with three cheers and marched on. As the miners approached, the Chinese all quit work, ran to their camps, grabbed their belongings, and took to the hills. In all about sixty Chinese were driven off. The miners marched out some three miles to the Storey County line, warning the foremen at each camp not to put the Chinese back to work. On the return march they leveled all the Chinese "shanties" in their course. The railroad suspended all construction work within the county.[24]

The Nevada and California dailies were unanimous in their condemnation of the "outrageous ... grovelling ... brutal ... guerrilla" action of the Miners' Union. Accusing the miners of the "lowest and meanest prejudices," they then indulged in

some of their own. The Sacramento *Union* editor felt that the
mob must have been "largely composed of foreigners, since native Americans are everywhere proverbial for their respect for
the law and the lawful rights of others." He was joined by a Carson City editor who speculated that the mob's leader was either
some "lately liberated slave from Ireland, a recently imported
Cornishman, or an imperfectly naturalized Canuck." A Republican editor attributed it all to the "encouragement" of the Democratic press and politicians.[25]

All called for the arrest and punishment of the participants.
But Joe Goodman of the *Territorial Enterprise* asked

should they be arraigned before the bar of justice, could a jury be
found in Storey County, unless especially selected for the purpose,
that would pronounce against them the verdict of guilty? We doubt
it. This is a strange condition of affairs, yet scarcely more to be wondered at than the circumstances which in a measure account for if
they do not justify it. They were made notorious daily by the mismanagement of our leading mines, of grasping monopolies, of a reckless waste of the substance of the commonwealth . . . and everything
looking toward the correction of the alleged evil, whether it take the
form of moral force or personal violence, is regarded with dangerous
favor.

No legal action was, in fact, ever taken against any of the miners,
and their spokesman, Thomas Atkinson, was subsequently
elected county sheriff.[26]

There was nonetheless condemnation of the affair from within
the Miners' Union and much confusion was generated in trying
to lay the blame. Some Gold Hill miners charged that only the
Virginia City union was responsible, while others claimed it was
only the work of a small minority. Finally the president of the
Virginia City union claimed the unions "had nothing to do
with it."[27]

Still the officers of both unions sought a settlement of the issue
with William Sharon when he returned from San Francisco a
week later. Sharon suddenly became a rousing Anti-Coolieite
himself, when he addressed the Gold Hill miners from atop a pile
of mining timbers at the Yellow Jacket on October 6th. Claiming he had been "vilified and maligned" by charges that "I am

in favor of introducing Chinese labor into these mines," he explained,

that is false. I never have had any such idea; neither have I now. The Chinese are no miners, and will not work underground where you do. We want *miners* and not Chinamen. They have no interest in the country, nor even religion, tastes and ideas in common with us, and we can only employ them in menial service, inferior occupations, railroad grading, and all that sort of thing. . . .When it is completed the Chinese are no longer wanted and can go.[28]

Sharon backed up these promises by signing a pledge "that I have no intention to place Chinese labor in the mines or mills, and that they shall not be worked on the railroad north of the American hoisting works." This was acceptable to the unions and work resumed two days later. The Chinese stayed south of the American works and the railroad was completed without further incident.[29]

This was the first and last anti-Chinese incident on the Comstock. The miners had reacted so strongly and so swiftly in barring even Chinese graders from the county and they were so solidly backed by the rest of the community that the mine managers apparently concluded that any attempt to employ Chinese in the mines and mills would be futile. Even if the mining companies could ultimately have succeeded, Sharon and the "mill ring" doubtless opposed any attempt since they personally would have gained little from a reduction in mining costs while they would have lost much in hauling and milling profits from the long and disruptive strike that would certainly have ensued. Confident in their success, the miners' unions turned their attention to other issues again and the anti-Chinese fever subsided for a time. The issue, in fact, seemed so well settled that the Grand Council of the State of Nevada never even bothered to meet again.

In California at this time the anti-Chinese fever was raging in the larger cities, but the Chinese in the placer mines were unaffected and there was only a brief flare-up in the deep mines. The employment of Chinese at Grass Valley had been suggested by one mine owner during the Giant Powder strike there in the

spring of 1869, but even the suggestion had been roundly denounced. Still the possibility prompted the organization of a
short-lived Grass Valley Anti-Coolie Association in September
"to prevent the employment of Chinese, as laborers, in any
branch of industry." This was superseded by the Order of White
Men's Defenders, which took part in a successful and inglorious
campaign to defeat a judge who had been so brash as to admit
the testimony of a Chinese witness in a local trial. On the Mother
Lode the Amador County Laborers' Association had sought
among other purposes "to discourage competition of inferior
races." But the employment of half a dozen Chinese mill hands
was only a minor issue in the Amador War of 1871. Only in the
deadly quicksilver mines in the Coast ranges were Chinese miners employed underground without opposition.[30]

Elsewhere in the West in the early 1870s there were only a
few isolated incidents of Chinese persecution by the miners, as
they moved quickly and forcibly to bar them from gaining any
work in the deep mines. During the summer of 1872, while many
miners around Silver City, Idaho, were out prospecting, labor
was scarce and several companies hired Chinese for ore sorting
and other surface work. No opposition was voiced until winter
found some of the returning prospectors unable to get jobs. By
February of 1873 there were nearly a hundred unemployed miners in the camp and the Fairview Miners' Union concluded to
take action, demanding that no Chinese be employed in or
around the mines as long as there was "white labor" available.
A list of men willing to work was given each superintendent.
But vowing that they would not be "dictated" to, the superintendents refused to discharge the Chinese. On Washington's
birthday the union decided to strike until the demand was met.
The following day all the miners on War Eagle Mountain walked
out; only the engineers on the pumps continued work.[31]

The editor of the local paper, the *Owyhee Avalanche*, denounced the action, but his concern was for the rights of the
mine owners, not the rights of the Chinese. For he warned that
the camp would "dwindle down to a one-horse concern unless
mine owners are permitted to manage their property as they
deem fit." One owner, in fact, swore, "I would rather see water

flow out of the shaft for the next six months, than accede to the demand of the Miners' Union." But most of the community supported the miners and within two weeks the superintendents agreed to the demand, fired the Chinese, and resumed work with all-white crews.[32]

Although there was no miners' union at Nederland, Colorado, the reaction of the miners to the introduction of Chinese there was also swift and summary. On March 29, 1874, when the Mining Company Nederland brought in 160 Chinese to work in their mine and mill, a band of forty masked and armed men forced them to leave before they even started work. The town board of trustees resolved that the guilty would be punished and that no such outrage would be permitted again. But the resolutions were doubtless intended only to impress outside capital, as no one was arrested and the company decided not to risk bringing in more Chinese.[33]

No further attempt was made to employ Chinese in any of the lode mines east of the Sierra. But some Chinese did settle in and around the lode mining camps where they were permitted, for a time at least, to work as cooks, laundrymen, and woodcutters. There they lived relatively free from harassment for a few years until hard times brought a new epidemic of anti-Chinese fever in the latter part of the 1870s. Then once again the Chinese were driven even from these jobs and often from their homes by mob edict. The epidemic began with the violent anti-Chinese demonstrations of Denis Kearney and the sandlotters of San Francisco who carried the Workingmen's Party of California to sufficient power to write discriminatory laws into the state constitution in 1879. As the epidemic spread it infected both major national parties and finally culminated in a renegotiation of the Burlingame Treaty in 1880, the passage of the Chinese Exclusion Act in 1882, and a frenzy of anti-Chinese riots throughout the West in 1885 and '86.

In the California mines, the new wave of anti-Chinese agitation even hit the Chinese in some of the placer mines—the large hydraulic companies working the placer gravels that stretched for nearly thirty miles along the Inter-Yuba Ridge from Smartsville to North Bloomfield. The agitation led to the establishment of the first miners' union in the placer mines. It began in March of 1876 with the liberal-sounding proposition that Chinese be paid "the same wages per diem as white men receive for the same kind of work." Anticipating that bigoted superintendents would fire the Chinese rather than pay them "white men's wages," the white miners at Sweetland and French Corral in the middle of the Ridge organized a Miners' League to try to enforce this demand.[34]

William Ewer of the *Mining and Scientific Press* thought this was "assuredly a queer way of treating the question," but granted it "the merit of novelty." The mine owners, taking a less amused view, rejected the demand. One superintendent, however, Judge V. G. Bell of the Milton Water and Mining Company, did offer a counterproposal. Although he claimed he could not afford to work the mines at higher wages, he was willing to discharge his Chinese miners and hire "white boys" at the same wages, getting them off the streets and making "good men" of them. But the miners, failing to appreciate the virtues of child labor, rejected the counterproposal.[35]

The League soon gave way to a more violent statewide group, the Order of Caucasians, headed by Supreme Constable Thomas Loyd of Grass Valley. In December of 1876 the miners along the Ridge organized a Camp of the order at North San Juan. The Caucasians declared all who patronized or favored Mongolian labor to be "public enemies" and they solemnly pledged "to impede, harass and destroy a public enemy by every mode and means and manner, known and unknown, within the reach of brains and thought and act and within the bounds of law."[36]

Still the Caucasians took no action until February of 1877 when several hydraulic mining companies announced a wage cut from $3.00 to $2.50 for white labor and from $1.50 to $1.30 for Chinese labor. The miners down at Smartsville struck to resist the cut and so did the Chinese miners up at North Bloomfield. But the Caucasians around French Corral agreed to the cut if the companies would discharge the Chinese and replace them with white miners at the reduced rate. Bell of the Milton company, employing 170 Chinese, accepted the proposition but the Caucasians were unable to find enough miners willing to work at the lower wages.[37]

Frustrated in these negotiations the more rabid Caucasians apparently turned to arson, setting fire to the Chinese quarters at French Corral, Grass Valley, Auburn, and Dutch Flat. Fortunately no lives were lost, but there was enormous property damage. The local press and sheriffs claimed that most of the fires owed to "carelessness" on the part of the Chinese, and only at French Corral was anyone charged with arson. These incidents, however, marked the end of the Order of Caucasians.[38]

Because of their lack of solidarity in opposing either the wage cut or Chinese labor, the miners along the Ridge had lost both issues in the spring of 1877. But during the next few years they finally sorted out their priorities and organized to reestablish the $3 minimum wage. Miners' unions were formed at Moore's Flat and North Bloomfield on the upper end of the Ridge, at French Corral in the middle, and at Smartsville on the lower end. Within five years they boasted a combined membership of roughly a thousand.[39]

These were the first labor unions organized in the placer mines and their appearance reflected the growing industrialization of even this phase of mining. Previously placer mining with pan, rocker, or sluice in the stream beds had been essentially an individual or small-group effort. But the hydraulicking of ancient river beds, such as the gravel deposits of the Inter-Yuba Ridge, required much larger capitalization to build the dams and flumes necessary to supply the great volumes of water used in washing out the gold. With increased industrialization, absentee ownership, and attendant dehumanization of the miner came the need for unionization. These unions were also modeled after the Comstock unions, but they added pledges in their constitutions "to resist the working of Chinese in the mines."

Once the unions had fairly well established $3 wages again, they called for the firing of all Chinese. Taking on the companies one at a time they demanded replacement of Chinese by white miners. The anti-Chinese feeling in California had grown so strong that their demands were met with little opposition.

Only in a few instances was there even brief resistance. At the Rainbow mine, when the miners' union called for the discharge of the Chinese, the owners fenced in the mine, armed the Chinese with Winchesters, and gave orders to "shoot the first man that intrudes." At the Blue Tent mine the Chinese armed themselves. But the local press and the rest of the community sided with the miners' unions and the companies soon capitulated without a fight. The unions tried to convince the companies that they could operate more economically with half as many white miners at $3 a day as they had with Chinese at $1.50 a day. But some who agreed to discharge their Chinese hands found even greater economy by hiring whites at only $2.50. By April of 1882 the Chinese had been discharged from all of the larger mines on the Ridge and a local editor noted that they had become so "unpopular" that soon there would be none left in any of the mines. But the white miners had little opportunity to enjoy their victory. The massive debris from the hydraulic mines so choked and polluted the streams that farmers in the Sacramento Valley below, finally secured permanent injunctions to close down the mines.[40]

The hardrock mining camps beyond the Sierra Nevada were also swept up in the new anti-Chinese frenzy. But here since there was no longer even a threat of Chinese employment in the mines, the miners' unions took no direct part. Although many of the miners were still active in the agitation, it had become a much broader issue, exciting the entire community. Anti-Chinese clubs, calling for complete expulsion of Chinese, thus sprang up independent of the miners' unions in many of the western camps ranging as far to the east as the Black Hills. But since most of the Chinese in the hardrock camps were employed either as cooks or laundrymen, the anti-Chinese agitators, for all their bombastic rhetoric, were usually unwilling to run them off till someone could be found to replace them. Only when they could find Chinese whose work was of no direct benefit to them, such as woodcutters for the mills and smelters, or graders for the railroads, did they actually try to enforce expulsion. Thus despite all the renewed agitation for Chinese expulsion, there were only a few incidents and these met with only temporary success if any.[41]

In April of 1876 the charcoal contractors for the Tybo Consolidated Mining Company in central Nevada brought in a dozen or more Chinese from Eureka to cut wood. A Workingmen's Protective Union was promptly formed to "permit no Chinese labor in the camp." In the middle of the night a "committee" of seventy escorted the Chinese out of town. The following morning the contractors sent out several men, armed with shotguns and Henry rifles, to bring them back. But the Workingmen's Union persisted. They raised $165 to hire wagons to take the Chinese all the way back to Eureka and they gave the contractors twenty-four hours to discharge them. The contractors stalled for a short time, but finally agreed when the mob of about 150 armed men threatened to expel them too. No legal action was taken against the union. In fact, one local editor complimented the Tybo miners on their "magnanimity" in providing wagons for the Chinese and he dismissed the expulsion as just "another argument in favor of the necessity of national action upon the Chinese question."[42]

An Anti-Chinese Club was organized at Eureka soon afterward and to prevent a similar dispute there the mining companies stipulated that their contractors not employ Chinese. In December the Club finally found a target, temporarily driving off the Chinese graders from the construction camp of the Eureka and Palisade railroad.[43]

The only other major attempts to expel Chinese from the hardrock camps during this period were more spontaneous affairs, lacking any formal organization. The first occurred at the silver camp of Bodie in eastern California, just across the line from Nevada. There in May of 1881 about four dozen Chinese began grading at the southern end of the Bodie Railway and Lumber Company's line, some thirty miles from the camp. A protest meeting was called, curiously the principal speaker Ben Butler, a lumber contractor, was much more concerned with being put out of business by the company than with the employment of Chinese. But when the railroad superintendent refused to discharge the Chinese immediately, the meeting resolved to "drive them out."[44]

Shortly after midnight some forty to fifty drunken raiders set out for the grading camp on the shore of Mono Lake. Some were carried by Butler's teams; some went on horseback and some by foot. But when they reached the camp late that day they found that the Chinese had taken the railroad company's steamboat to Paoha Island in the middle of the lake. There they were safely encamped, feasting on gull and duck eggs cooked in the island's hot springs. As the only available boats were two rickety little skiffs, the frustrated raiders could do nothing but "stand on the shore and make faces." A buckboard loaded with whiskey and provisions for a siege was sent down to the lake, probably by Butler, but it failed to connect with the raiders who came straggling back into Bodie the following day.[45]

The editor of the Bodie *Free Press* dismissed the raiders as "ignorant and weakminded laboringmen—egged on by a few bankrupted wood men," and viewed the whole incident as comic opera. The Bodie Miners' Union, which had refused to endorse the raid, won praise even from the antilabor San Francisco *Stock Report* for having "shown capitalists that enterprising undertakings . . . will be fostered and protected by the Union."[46]

Similarly the miners' union at Butte, Montana, remained aloof from agitation against Chinese woodcutters there later that same year. Gong Wung Lung and Company took a contract to cut wood for the Colorado and Montana smelter in December and put a crew of forty Chinese to work a dozen miles east of town. This sparked anti-Chinese agitation, culminating in a raid on the woodcutters. An armed mob of more than two hundred drove the Chinese from the timber shortly before Christmas. But after several of the leaders were arrested, the Chinese returned to complete the contract without further incident.[47]

These efforts at expulsion of the Chinese seem to have been only exercises in frustration, but they sharply exposed the violent antagonism that the hardrock miners and much of the community felt toward the Chinese. Although the miners' unions, seeing no direct challenge, refused to back such efforts, the miners themselves were vigorous participants. Thus even if the raids were ineffective in expelling Chinese from the periphery of the mining industry, they nonetheless revealed the depth of opposition that would confront any move to give the Chinese a more important place in the industry, and they stood as a clear warning against such a move.

But the strongest impact of the anti-Chinese agitation during the late 1870s and early 1880s was felt in the political arena. Here the hardrock miners and their unions played only a small role. Beginning in 1877 Denis Kearney's Workingmen's Party of California consolidated the anti-Chinese movement throughout the state into a successful political machine which elected mayors, assemblymen, and state senators, wrote anti-Chinese legislation into the new state constitution of 1879, and finally forced both the Republican and Democratic parties into supporting a renegotiation of the treaty with China in 1880 and the passage of the Chinese Exclusion Act in 1882.[48]

In the mines a combined "Miners' and Workingmen's ticket" won seats in the state Constitutional Convention from Nevada County and John A. Eagon of the Amador County Laborers' Association won a nonpartisan seat from Amador County. Led by the Workingmen's party delegates, the Constitutional Convention adopted a number of provisions discriminating against the Chinese. They were ratified by the people of California on May 7, 1879. The provision of most interest to the miners was Article 19 Section 2, which prohibited all corporations within the state from employing Chinese "directly or indirectly in any capacity." To enforce this provision the legislature made employment of Chinese a misdemeanor and imposed stiff penalties on the offending corporation and its officers. Some companies discharged their Chinese employees in compliance with this law but the federal courts soon declared it and other anti-Chinese provisions to be in violation of both the Burlingame Treaty and the Fourteenth Amendment.[49]

On the national level at about the same time Congress passed a bill limiting the number of Chinese immigrants to fifteen per vessel, but President Hayes vetoed this measure also as a violation of the treaty. These decisions spurred increased demands for the abrogation of the Burlingame Treaty and the govern-

ment finally acceded. On November 17, 1880, a new treaty was
signed with China permitting "reasonable" regulation, limitation, or suspension of immigration of Chinese laborers, but not
absolute prohibition. In March of 1882 the Congress passed a
bill suspending Chinese immigration for twenty years, but President Arthur vetoed this measure as an "unreasonable" restric-
tion. Immediate indignation over this action was expressed
throughout the western states. Flags were flown at half-mast on
the Comstock and the miners at Sutter Creek strung up effigies
of the president while the band played the dead march. As
pressure for a compromise mounted, a bill excluding Chinese
laborers for ten years was finally signed by the president on
May 6, 1882. In 1892 the period of exclusion was extended for
another ten years and in 1902 it was extended indefinitely.[50]

With the passage of the Chinese Exclusion Act, anti-Chinese
agitation shifted toward the expulsion of Chinese already in the
country. Even before the president signed the Exclusion Act,
the San Francisco Trades' Assembly called a convention of
western labor unions and anti-Chinese organizations to devise
a way to "settle this matter now and forever." The hardrock
miners' unions were well represented. Some forty-two miners
from the Gold Hill, Virginia City, Ruby Hill, and Cherry Creek
unions in Nevada and the Moore's Flat and North Bloomfield
unions in California made up a fourth of the delegates to the
convention.[51]

They met in San Francisco on April 24th, 1882, with Robert Morrison of the Virginia City Miners' Union as chairman.
After much oratory the delegates decided to form a new organization, the League of Deliverance. They dismissed the Exclusion Act as wholly inadequate, arguing "even absolute prohibition is not enough to destroy this wrong. PROHIBITION AND
EXPULSION are the true remedies."[52]

But the passage of the Exclusion Act seemed to have calmed
the anti-Chinese fervor of most westerners, at least for a time.
Attempts to establish branches of the League in the mines and
elsewhere met with little success and it disbanded within the
year. Its demise marked the last significant involvement of the
hardrock miners' unions in the anti-Chinese movement.[53]

Still anti-Chinese agitation flared anew a few years later when other western laborers came to realize that the Exclusion Act had not magically ended Chinese competition. The policies of expulsion outlined by the League were then put into practice in various parts of California, the Pacific Northwest, and the Rocky Mountains with violent results. The bloodiest incident occurred in the coal mines at Rock Springs, Wyoming, where Chinese miners had been employed for several years. There on September 2, 1885, twenty-eight Chinese were savagely slaughtered and hundreds were driven from their homes by a blood-thirsty mob, nominally led by the local Knights of Labor. Even after such violent outbreaks subsided, boycotts of Chinese laundries and restaurants continued sporadically for another decade or more.[54]

But by the late 1870s the Chinese question had become increasingly an issue of national concern as Chinese began to find employment in more varied occupations and locales. At the same time the hardrock miners and their unions became less concerned with the issue, primarily because their earlier militant actions in barring Chinese from all work in and around the deep mines had proved so decisive that the possibility of Chinese competition no longer seemed a real threat to them.

The Bonanza Years

The years from the early 1870s through the early 1880s were bonanza years of prosperity and growth for both the mining industry and the mining labor movement in the West. The first half of this decade saw the opening of the fabulous riches of the Big Bonanza on the Comstock Lode and the strengthening of the Nevada miners' unions under the leadership of the Comstock unions. Although the later years witnessed the exhaustion of these bonanza ores, the decline of the Comstock was offset by the rise of rich new camps—Bodie, Deadwood, Butte, Leadville, and scores of others throughout the West. With the decline of the Comstock many miners joined the rushes for the new boom camps, carrying the spirit of the union movement with them. New miners' unions were formed in nearly all the major camps and in many of the smaller neighboring camps as well. In all more than three dozen new miners' unions were organized by the mid-1880s, increasing the number of hardrock unions fivefold and firmly establishing the mining labor movement throughout the West. The only major defeat of the movement during this period was the curiously anomalous strike at Leadville in 1880.

Throughout this decade the Comstock miners' unions remained the largest and strongest, and the bastion of the western mining labor movement. Under the leadership of the Comstock unions all the miners' unions in Nevada were "affiliated and consolidated" in the summer of 1877, the first such confederation of western miners' unions. The exact nature of the affiliation, however, was not made public. Initially it included the Gold Hill, Virginia City, Silver City, Austin, and Ruby Hill miners' unions. Within the next few years several new unions, formed elsewhere in the state, joined them. Their combined membership numbered more than four thousand. Under this affiliation members of one miners' union were "equally members of all other unions in the State." Thus a miner moving from one union camp to another could join the union there without the trouble and expense of a second initiation. These arrangements were formalized in amendments to the union constitutions, establishing Clearance Certificates, which entitled transferees "to the same privileges and benefits . . . as those who came into the union by initiation."[1]

But the main goal of affiliation was solidarity, as it was "their fixed purpose to stand or fall in a body." Thus the first concerted action undertaken by the affiliation was the establishment of a closed shop in all the union camps in the state. The Ruby Hill Miners' Union had been unsuccessful in trying to get a closed shop in the spring of 1876, when the superintendent of the Richmond Consolidated Mining Company began discharging union miners in an attempt to break the union. The Comstock unions on the other hand had refrained from seeking a closed shop in view of the deathblow that issue had dealt the former Miners' League. By 1877, however, the miners' unions on the Comstock and in the other major camps were solidly established and most of the working miners in these camps were already union members. By then too the union miners were nearly unanimous in believing that, since all men underground enjoyed the high wages maintained by the union, "all should equally share its responsibilities."[2]

Immediately after their affiliation in the summer of 1877 each of the unions demanded a closed shop. The Virginia City, Gold Hill, and Silver City unions announced that none but union men would be allowed to work underground after September 10th. The Ruby Hill union notified all nonunion men to join the union by September 15th and the Austin union set a similar deadline. The unions also demanded that new miners join the union within thirty days after they start work. The demands met considerable editorial opposition, particularly in the San Francisco papers, but in the mining camps they received strong endorsement. As the editor of the Eureka *Sentinel* argued,

It is a fact, admitting of no doubt, that to the miners' unions of the State belong the credit of keeping wages at four dollars per day for all underground work. . . . Nor is this all. There is every probability that had not these protective associations been formed, a very considerable portion of the mines of Nevada would be to-day worked by Chinese labor, to the serious detriment of all other classes and interests. Non-union men have for years enjoyed the advantages secured to members of the Union, while bearing no part of the burdens and responsibilities it entails. It is therefore right and proper that they should be invited to throw their weight and influence in favor of the maintenance of living wages, which they have so long been able to demand and receive through the efforts of others.[3]

The Comstock unions won their demand without opposition, but at Austin and Ruby Hill some nonunion men failed to sign up. The unions there postponed the deadlines until November. By the new deadline all but one shift boss in the Manhattan company's mines at Austin had joined. The miners, however, demanding strict compliance, struck and within hours the superintendent had convinced the last holdout to sign up. Thus the closed shop was established in all the major camps in Nevada. As new unions formed and affiliated with the older unions, they also established closed shops after they reached sufficient strength to back such demands.[4]

But the exhaustion of the Big Bonanza ores and the resulting cutbacks in the Comstock mines also began in 1877. During the next two years Comstock bullion production fell from $37 million to $7.5 million a year. The principal mines started cutting back their work in 1877 and by the end of that year crowds of unemployed miners gathered at the various hoisting works at every change of shift hoping to fill the place of some man killed or discharged. Many of the single miners soon moved on to more promising camps. Within the next four years the Comstock companies had cut their work forces roughly in half before the retrenchments ended. Nearly two thousand miners had left the Comstock, spreading throughout the West wherever new mines were opening, carrying the union movement with them.[5]

These miners, leaving the Comstock in the late 1870s, were not like the free-footed rainbow chasers, who rushed to every new boom. They were looking not for a quick fortune but for a steady job at good pay. Their years of experience on the Comstock had made them highly skilled miners who could demand and get a good wage, but they also fully realized the need for a solid union to maintain it. Thus despite the unsettled nature of many of the boom camps in which they landed, they hung solidly together to build strong new unions.

Some of the first miners to leave the Comstock in the fall of 1877 joined the rush south to the silver camp of Bodie, just across the line in California. That winter, on December 22, a dozen former Comstock miners met in Williamson's saloon to form the Bodie Miners' Union. They adopted the constitution and bylaws of the Virginia City union and Alex Nixon, a former member of both the Virginia City and Ruby Hill unions, was elected president. There was no threat to the $4 wage at that time and with the strong union spirit of the Comstock miners the Bodie union grew rapidly with the camp. Within six months the miners had a Union Hall, which still stands today. By May of 1880, when they declared a closed shop, they numbered over a thousand members. The Bodie Miners' Union for a time was the strongest in California and it continued into the present century. Under its patronage miners' unions were organized in the neighboring camps of Lundy in February of 1881 and Aurora in March of 1882.[6]

Comstock miners returning to the older gold camps of California spread the union movement in Nevada county and along the Mother Lode, organizing unions at Moore's Flat and Columbia in 1877, Nevada City in 1879, Forest City in 1881, Placerville in 1882, and Alleghany in 1883. At the same time the hydraulic miners along the Inter-Yuba Ridge were caught up in the movement, forming unions in the late 1870s at Smartsville, French Corral, and North Bloomfield, both to resist a wage cut and to drive out the Chinese.[7]

Other Comstock miners joined the rush to the Black Hills of the Dakotas which had begun in 1876. There they helped to organize miners' unions at Lead and Central City in October and November of 1877. These unions also adopted almost verbatim the constitution and bylaws of the Virginia City union. Moreover the Lead union sent its first president, J. T. Tully, on a 3,000-mile round trip to the Comstock to learn from the officers of the Virginia City union the subtleties of union policy and procedure. Competition of cheaper labor from the Midwest drove wages down and the Black Hills unions struggled for nearly a year before they successfully established a $3.50-a-day minimum wage. But thereafter they grew rapidly. By 1882 the Lead union had 850 members and the Central union only slightly fewer. These unions are still active today. The miners at nearby Bear Butte organized a union in September of 1879 and those at Terry's Peak organized a few years later.[8]

At the same time Comstock miners drifted into the Montana silver camps. When the owners of the Alice and Lexington mines tried to cut wages from $3.50 to $3 a day, the miners struck and formed the Butte Workingmen's Union on June 13, 1878. This day is still celebrated as Miners' Union Day in Butte. The Butte miners adopted the constitution and bylaws of the Comstock unions and soon reestablished a $3.50 minimum wage. But the union grew slowly until the commencement of copper mining in the mid-1880s. Renamed the Butte Miners' Union, it then became the largest and most powerful in the West, surpassing even the Comstock unions. It boasted more than 4,600 members by 1893 and was the prime force behind the formation of the Western Federation of Miners that year. In October of 1878 the miners organized a union at Phillipsburg, which at that time was the only other major camp in the territory.[9]

The rush to Leadville, Colorado, which began in late 1877, also attracted many miners from the Comstock, but the camp drew the bulk of its population from the Midwest and East. In January of 1879 the miners there formed a conservative Miners' Cooperative Union as an assembly of the Knights of Labor. The following year it was reorganized along the more militant lines of the Comstock unions. But it was soon broken by the intervention of the National Guard in a bitter strike that left Leadville the only major nonunion camp in the West.[10]

Comstock miners also joined scores of smaller rushes in the late 1870s and early 1880s; to Utah where miners' unions were organized at Silver Reef and Park City in February and June of 1880; to the Wood River country of Idaho where the Bullion, Broadford, and Ketchum unions were established in 1881, 1882, and 1885; to the San Juan country of southwestern Colorado where the Telluride and Silverton unions were formed in the early 1880s; to the new camps in central and northern Nevada where unions were organized at Cherry Creek in November of 1879, at Tuscarora and Grantsville in June of 1880, at Battle Mountain and Lewis in February of 1881, at Spring City about 1883, and at Candelaria in 1885; and finally into southern Arizona where unions were formed at Tombstone in April of 1884 and at Globe and Bisbee a month later.[11]

Thus in little over half a dozen years starting in 1877 more than three dozen miners' unions were organized throughout the West, representing nearly ten thousand miners and greatly enlarging the base of the mining labor movement. And throughout this period, despite the decline of mining on the Comstock, the Comstock unions continued to dominate the movement. Almost without exception the new unions adopted the constitution and bylaws of either the Virginia City or the Gold Hill union with little more revision than altering the meeting day or raising the sick benefit a few dollars to cover higher costs. Most of the new unions also established ties with the Comstock unions, either through affiliation or through personal contact and correspondence. A number of the new unions were, in fact, spoken of as "branches" of the Comstock union, but the relationship of those outside Nevada was undoubtedly only informal.

The new unions also followed the pattern of the Comstock unions in benevolent, social, and political activities. The bulk of their energies and resources were spent in aiding the sick, the injured, and the bereaved. In addition they built union halls that were frequently the largest meeting place in the camp, serving not only for their weekly meetings and occasional fund-raising balls but also for meetings of the Masons, Odd Fellows, Hibernians, and other fraternal groups, for performances of local and itinerant lecturers, singers, and players, and for a variety of charity benefits. In addition the miners' union hall served other community functions, ranging from town hall and polling place to library, school, and church. In the political arena the miners' unions frequently elected one of their members justice of the peace or sheriff and occasionally state or territorial legislator.

Finally nearly all the new unions followed the Comstock unions in making the establishment of a minimum wage their principal goal. Most were organized before wage cuts were imminent and they simply adopted the prevailing wage in the camp as their avowed minimum wage. A few of the unions, however, were formed to resist a threatened wage cut and they were generally successful. In the silver camps from the crest of the Sierra Nevada to the crest of the Rocky Mountains—in eastern California, Nevada, Idaho, Utah, and western Colorado—all the new unions set the minimum wage at the Comstock rate of $4 a day. Within these mountain walls the power of the Comstock unions was unchallenged and all pervading. But beyond the crests the cheaper labor markets of the Midwest and Pacific Coast forced a lower wage. From the crest of the Rockies east—in the copper camps of Montana and the gold camps of South Dakota—the new unions had to compromise at $3.50 a day, and even this was won only after prolonged strikes at Butte and the Black Hills. Still it was well above the $3 or less a day earned in the ununionized silver-lead and gold camps on the eastern slopes in Colorado. Similarly on the western slope of the Sierra in the gold camps of California the unions had to settle for $3 a day and the hydraulic miners had to win a long series of strikes and drive out the Chinese even to get that.

During these years the Nevada unions faced only minor encroachments on the $4-a-day minimum wage, spurred primarily by the gradual decline of silver prices. Following demonetization in 1873, silver fell from its monetary value of $1.33 an ounce to about $1.15 an ounce in 1876, before it stabilized for a time. The decline had surprisingly little effect on the western silver mining industry as a whole and production continued to rise as new ore bodies were discovered and more efficient technology was introduced. Still, individual mines, which had been only marginally profitable, did feel the pinch, and inevitably a few looked to some form of wage cut as a remedy. In the spring of 1875, when the price of silver had slipped well below its monetary par, Allen Curtis, agent of the Manhattan Silver Mining Company in Austin, announced that after May 31st he would pay all hands in "bullion checks," or bars of "merchantable silver at par." He proposed to strike the bars in denominations of $3, $5, $10, $20, $50, and $100, and ordered dies from San Francisco.

The general store in which Curtis was a copartner advertised that they would accept the checks at par but most other merchants indicated they would discount them 5 or more percent to the market value. Thus on May 31st the miners' union voted to demand their wages in gold coin, and to strike if Curtis refused. Curtis, not wanting "obstinancy" to injure the company, abandoned the plan.[12]

Several months later the Ruby Hill union resisted a similar attempt by the Richmond Consolidated Mining Company to pay its miners in depreciated silver trade dollars. But the victory there was short-lived and only opened the way to new abuses. The superintendent, R. Rickards, promptly discharged all the men on daily wages and let the work out on tribute or contract. The miners called a strike. Union president Henry Ratcliffe, denouncing the new system as "rotten and corrupt," charged that the contracts were not open to competition but were given out secretly at exhorbitant prices to the superintendent's friends and relatives, who then subcontracted the work to miners at much less. Although much of the community had applauded the union's stand against trade dollars, many now felt that the union had gone too far. They argued that as long as the miners hired by the contractors received full wages, the union should let the stockholders worry about abuses of the system.[13]

The union, however, continued its strike against both the company and the contractors, and soon all work was suspended. As passions flared one contractor received an anonymous note, depicting a coffin with his initials on it and "threatening him with the contents of a rifle barrel." The threat was disclaimed by the union but the contractor left for Virginia City anyway. After several weeks the issue finally seems to have been compromised informally, as the Richmond resumed work with some men on wages and some on contract.[14]

One wage-cutting scheme was even tried on the Comstock when it first began to decline. In April of 1877 a would-be labor broker, J. D. Bodwell, operating under the fictitious title of the Co-operative Milling and Mining Company, offered to secure jobs for miners at $4 a day in exchange for "their notes for one hundred dollars, payable in installments of one dollar per day."

The Comstock unions denounced the scheme, notifying him to "desist." But when a group of miners later found him trying to recruit more men, they began to heckle him and one man, pressing through the crowd, yelled that he had been swindled. At this Bodwell dashed through a tin shop and fled down the canyon with the crowd in hot pursuit. He never came back.[15]

Even when the miners were promised good pay they did not always receive it. The miners of Silver City, Idaho, were frequently victimized by defaulting companies. "The past mismanagement of affairs in this camp," the local editor noted, "has told with crushing effect upon the miner, who deprived of his hard earnings from time to time, has walked these hills with empty pockets, save in the possession of valueless time checks, not knowing often where his next meal was to come from. Many poor fellows left the camp in despair with their blankets on their backs and not even a small portion of their just dues in hand."[16]

But the miners finally took action on June 30, 1876, when the principal mine, the Golden Chariot, closed down, owing its crew nearly two months' back pay. It was rumored that M. A. Baldwin, the superintendent, was planning to leave that same evening for San Francisco with four to five thousand dollars worth of quicksilver. So the desperate miners moved swiftly. Taking Baldwin hostage, they demanded their back pay before they would release him. The sheriff visited Baldwin but made no effort to free him. The miners' union also unofficially endorsed the action and the local paper decried the treatment the miners had received "at the hands of 'soulless corporations' and worthless superintendents." The company directors in San Francisco promised to send the money but it never came. After three weeks of waiting the miners finally agreed to let Baldwin go to San Francisco to personally try to raise money, both to pay back wages and to resume operations. In the meantime the miners' union helped raise a relief fund of $1,100 to tide over the suffering miners' families. Two months later Baldwin returned with funds to pay off those miners who had remained and to reopen the mine.[17]

With the blessing of the local unions a much more direct

solution was tried elsewhere. In Phillipsburg, Montana, when the Northwestern company defaulted on the back pay of their miners in 1879, the miners seized the mine and worked it themselves till they had recouped their losses. Two years before Deadwood miners had been thwarted in a similar attempt at the Keets mine, when the sheriff called in a Cavalry company to smoke them out with asafetida.[18]

Following the precedent set by the Comstock unions in the dispute between the Justice and Alta mines, the miners' unions in other camps also intervened when lawless mining disputes threatened the lives of miners. The most decisive and controversial action was taken by the Bodie Miners' Union in drumming George Daly, and several of his henchmen out of the camp, after they had killed a miner in an adjacent mine. George Daly was, in fact, a "heartily hated," quarrelsome, little man who drew a gun at the slightest excuse and made enemies as easily. The more generous of his enemies dismissed him as a "strutting, cheeky, little blowhard, always fomenting discord." But the depth of feeling against him was plumbed by an acquaintance who contended "he has always been a creeping, crawling, bung-sucking sycophant, and has wriggled his way into positions which he was totally incompetent to fill. He knows absolutely nothing about practical mining, and less, if possible, of the proprieties and decencies of good citizenship." Still his friends praised him as "a very wide-awake and enterprising man, a staunch and true friend, and a bitter but not unreasonable enemy." The son of an Australian clergyman, Daly had worked as a scab printer on the *Alta California* in San Francisco. He then went to the Comstock as foreman and later proprietor of the *Enterprise* job shop. He became an agent of stock shark James R. Keene, and in 1878 with no experience in mining he was made superintendent of the Jupiter and Mono mines in Bodie and the old Real del Monte in neighboring Aurora.[19]

In August of 1879 five miners who owned the Owyhee claim adjacent to the Jupiter started sinking a prospecting shaft on the north end of their claim. Daly, contending that they were on Jupiter ground, told them to get off but they refused. Rather than bring suit against them, Daly quietly armed several of his men for an attack. After the Owyhee miners quit work on the evening of the 22d Daly's men led by his foreman, Joe Mc-Donald, seized the Owyhee shaft and set up a prefabricated cabin over it. Throughout the night sporadic shots were exchanged between Daly's men and the Owyhee miners, who remained in their own cabin several hundred feet to the south. Just at daybreak Daly's men rushed the Owyhee cabin, killing one of the miners, John Goff, and wounding two others. The Owyhee miners surrendered at gunpoint and Daly's men marched them into town. Joe McDonald tried to get the sheriff to arrest them for trespassing but he refused.

The town was seething with excitement. The miners' union met to discuss the situation and decided to take possession of the disputed ground in an attempt to forestall further violence. Several hundred strong they marched up to the Owyhee shaft, removed Daly's cabin, rolled it down the hill, and burned it. The editor of the Bodie *News* praised the miners' union, noting "in cases of dispute and threatened trouble, they are the great conservators of the peace, and are more to be relied upon for the maintenance of order in any matters affecting the mining interests and peace of the community than all other powers combined." At the same time feeling against Daly had become intense. He and several of his men were arrested for murder by the sheriff and taken under heavy guard to the county seat at Bridgeport.[20]

But the following month Daly and his men were acquitted by a grand jury and they returned defiantly to Bodie. The miners, denouncing the proceedings as a "judicial farce" by a packed and corrupt jury, were determined that Daly should not remain. On Saturday night, September 20th, the Bodie Miners' Union voted 500 to 3 to order George Daly, Joe and Barney McDonald, and four other gunmen out of the district within twelve hours.

Daly vowed to resist this "obnoxious and violent resolution," calling upon "all good and true men" for assistance. At the same time, however, he hired a few dozen more gunmen and barricaded himself at the Jupiter hoisting works. As the deadline drew near a citizens' committee was formed to mediate between Daly and the union. The committee was headed by attorney

Robert D. Ferguson, former president of Virginia City Miners' Protective Association. Through their influence Daly finally agreed that he and his men would turn over their arms to the sheriff and leave the camp within two days, on the condition that he be allowed to return for twenty-four hours or less at any time. They left the following evening.[21]

Although the miners' action had the support of most of the local community, press opinion elsewhere was sharply divided. The *Territorial Enterprise* denounced Daly's banishment as "the most infamous act that any band of men ever perpetrated on this coast," and urged, "if the mine managers can afford it, their duty is to close down their mines and keep them closed until this crowd is starved out of town. They are not worthy, in a free country like the United States, to be given employment."[22]

The editor of the Eureka *Sentinel*, on the other hand, argued that "the Bodie Miners' Union simply met force with force. The peaceable, law-abiding miners of that district recognized in Mr. Daly a dangerous man, and after bearing with him until his conduct became unbearable, they ordered him to quit the place at once and forever." He also pointed out that ironically many of the papers denouncing the action of the Bodie Miners' Union had endorsed similar acts by the vigilance committees in San Francisco and Virginia City.[23]

The action of the Bodie union was also strongly endorsed by the Nevada miners' unions. The Virginia City, Gold Hill, and Silver City unions passed a joint resolution, condemning the *Territorial Enterprise*'s criticism of the Bodie miners and congratulating them for "ordering the assassins and their murderous leader from your midst . . . preventing what threatened to be a bloody riot, a useless sacrifice of life and property, and an injury to the camp." The Austin and other unions followed suit.[24]

Daly, resigning his superintendency of the Bodie and Aurora mines, departed for the East to try to convince investors there to withdraw support from the Bodie mines. It was an idle threat. Instead he found that his reputation as a "fighter" and a foe of miners' unions landed him a position in the boom camp of Leadville. He brought his Bodie gunmen with him and there they gained even further notoriety.[25]

The Bodie Miners' Union subsequently intervened in other disputes between the University and Maryland mines in June of 1880 and the Standard and Bulwer in May of 1888. But, learning from the Daly incident, they handled these disputes much more conservatively, stopping short of banishing offenders. Similarly miners' unions elsewhere were also more conservative, though no less decisive, in their intervention in mining disputes and they generally won praise as "peacemakers." At Eureka, for example, when stinkpots and dynamited bulkheads threatened the health and lives of miners during a dispute between the Albion and Richmond companies in February of 1881, the Ruby Hill Miners' Union simply took all its men out of both mines and demanded that the companies stay out of the disputed ground until the matter was settled in the courts.[26]

As an outgrowth of such disputes the Nevada miners' unions had a bill introduced in the state legislature in 1881 to prohibit the employment of "fighters" in the mines. The bill made it "unlawful for any person, company or corporation to employ, cause to be employed, or permit to be employed, any person, or persons commonly known as 'fighters,' either for the purpose of retaining possession of any mine or other property, or for the purpose of taking, or attempting to take possession of the same." Anyone either carrying firearms or any deadly weapon, or knowing that force, intimidation or violence would be resorted to as part of his job would be considered a "fighter." Both he and his employers would be liable to a fine of $100 to $500, or imprisonment in the County Jail for one to six months, or both. Moreover the bill made it a felony, punishable by one to twenty years in the state prison for any such fighter to assault or even threaten anyone and, if he killed anyone, he would be deemed guilty of first degree murder and his employers would be punished as accessories. The mining companies were not ready for this kind of justice, however, and they killed the bill in committee.[27]

Following the organization of miners' unions in the new camps, the mechanics, engineers, and others also organized. But, although the Comstock miners' unions had lent their support to the demands of the Storey County mechanics' union, the miners' unions in some of the new camps did not always show the same solidarity, especially if the action of the other union affected them adversely. The Bodie Miners' Union, in fact, interceded on the side of the mine owners in February of 1879, when the Bodie Mechanics' Union struck for $5 a day and eight-hour shifts for engineers. The strike of about fifty engineers had closed down most of the mines, throwing nearly a thousand miners out of work. The miners' union condemned the mechanics for not giving the superintendents, and apparently the miners too, "proper notice of their intentions." The union thus pledged "to sustain the superintendents in reopening the mines, and to defend them in employing such engineers as they might choose to select." With this the strike was quickly broken.[28]

More often the miners' unions simply refrained from taking any official action. At Cherry Creek the miners' union refused to support a strike of the Mechanics' and Millmen's Protective Union for a minimum wage and a Citizens' Committee induced the union to rescind its demand. Similarly the Ruby Hill Miners' Union withheld support for a strike by the Eureka Coal Burners' Protective Association and the state militia broke the strike after a skirmish in which five charcoal burners were killed.[29]

During this period a variety of quasi-labor unions were also formed by miners in some of the smaller camps. Some, like the Tybo Workingmen's Protective Union, were intent only on driving out Chinese competition. Others seem to have started out as social, or perhaps political, clubs before they dabbled in labor problems. A strike in eastern California by the Workingmen's Club of Darwin against the New Coso Mining Company led to tragedy. Early in 1878 the smelter workers struck against a pay cut from $4 to $3 a day. In May the company tried to resume work with scabs at the new wage. The workingmen blocked the trail leading to the works, and the superintendent called in the deputy sheriff, the constable, and two deputized gunmen to escort the scabs. When the workingmen refused to disperse, the sheriff's men opened fire. One of the strikers, C. M. Delahanty, was killed by the first shotgun blast and the others scattered. Violence flared again that evening when the constable killed another striker in an argument over the first killing. Two days later, after a citizens' meeting praised the peace officers and denounced the Workingmen's Club for leading Delahanty to his death, the club disbanded.[30]

Still other organizations, like those that flourished briefly in the Colorado camps, sought only mutual protection for miners against claim jumpers. Nonetheless they dubbed themselves a Miners' Union at Red Cliff and Buckskin, a Miners' Protective Association at Alpine and Aspen, and a Miners' and Merchants' Co-operative Association at Pitkin, most of which sounded in name at least like labor organizations. Their constitutions and bylaws, however, were a curious blend of protective society goals and mining district law, reminiscent of earlier times when districts took upon themselves the role of law enforcement.

Membership was open to any person in any way "connected with a mine" and willing to pay a fee of 25¢ or 50¢. The Red Cliff Miners' Union, formed January 10, 1880, was apparently the first of the breed. Lamenting the trouble and litigation caused by claim jumping and other fraudulent means of gaining title to a claim, the union took it upon itself to protect claims, mediate disputes, and mete out punishment. Thus in addition to the usual officers, the union elected a marshal to serve in place of a town constable and executive committee to meet weekly, serving as a court "to hear and decide upon any question brought before it touching the rights of members." The committee's decisions were final unless appealed within thirty-six hours to a jury appointed by the union president. No professional lawyer was allowed to plead a case before the committee unless it was his own. Anyone convicted of claim jumping was ordered to leave the district within twenty-four hours and the marshal was empowered to call up the entire membership of the union to enforce the order. The Alpine Miners' Association had similar eviction procedures, but the other organizations considered this too severe, since many disputes arose from claims that overlapped out of carelessness rather than intentional fraud. Their members simply pledged to submit disputes to arbitration by a committee of three and abide by their decision. Each party to the dispute chose one arbitrator and these two chose the third. The advent of effective law enforcement in these camps soon brought the disbanding of these organizations.[31]

The only major setback to the great expansion of the western mining labor movement during this period was the breaking up of the miners' union at Leadville, Colorado, in the spring of 1880. Leadville had just surpassed the declining Comstock in annual bullion output and the breakup left the most productive camp in the West without a miners' union. The causes of the breakup and the strike that led to it were complex. At the time R. G. Dill, editor of the Leadville *Herald*, wagered that they "will probably never be fully understood." And indeed the Leadville strike remains one of the least understood, and yet one of the most discussed, incidents in the history of the western mining labor movement. Nonetheless the major cause of the strike is clear. "In the light of later events" even Dill concluded "it is now scarcely doubted that the strike was organized rather by certain mine managers than by the miners themselves, and for the purpose of covering up the poverty of some of the mines until the principal stockholders could unload." Thus the strike was a sad hoax, not only on the western miners but also on a large number of eastern investors.[32]

To understand the strike we must understand the affairs of the Chrysolite Silver Mining Company. This company, floated by George D. Roberts in New York at a paper capital of $10 million, had purchased the Chrysolite and other Leadville mines in October of 1879 from Colorado's Lieutenant Governor Horace A. W. Tabor and William Borden for $2,778,800, payable in stock and earnings from the mine. By April of 1880 when the stock was first offered to the public it appeared to be the safest and yet most promising in Leadville. In just five months the Chrysolite mines had paid exactly $1 million in dividends—over one-third the purchase price—and the manager claimed to have opened up several tons of new ore for every one mined. Still, even to the novice the stock seemed overpriced at its $50 a share par. The best offers were half that, and that price slowly declined. But considering the recent jobbery of other Leadville stock and the questionable credentials of its promoter, George Roberts, even this might be considered high. All Leadville stocks had gotten a black eye with the collapse of the Little Pittsburg stock in February of 1880, when it was discovered that the mine had been gutted to pay dividends to its promoters, Senator J. B. Chaffee and D. H. Moffat, and then unloaded at a high price on an unsuspecting public. George Roberts, a former San Francisco stock shark, best known for his part in floating the infamous diamond swindle in 1872, had, in fact, embarked on the same exploitive scheme in the Chrysolite and a few other Leadville mines he was promoting.[33]

Thus the Chrysolite too was nearly gutted, but the fact was skillfully concealed by its manager, Winfield Scott Keyes, and its ore deposits still appeared to be quite extensive. As veteran mining engineer T. A. Rickard later explained, in the Chrysolite

lead-silver ore replaced limestone, working out from certain almost invisible straight and regular fractures vertical and intersecting one another at right angles; and the mineralization was confined to a certain limestone bed about ten feet thick. Drifts had been driven along the center of each of the principal fractures, and the natural ore thus stoped out; but the walls were still ore, though of lower grade. The mine was thus developed into blocks by the intersecting drifts, as if all the bed were ore and had been systematically blocked out as a scheme of development.

By May of 1880 little more than a venire of ore remained. Keyes had to cut ore shipments and the company skipped its first dividend. Chrysolite stock dropped sharply to $13 a share. But Roberts, blaming bad roads for the lower shipments, confidently called in Rossiter W. Raymond, the eminent mining engineer and editor of the *Engineering and Mining Journal*, to personally appraise the value of the mine.[34]

Raymond, trusting Keyes, who had been a fellow student with him at Freiburg, was deceived into believing that the ground blocked out was all ore, when, in fact, it was nearly all barren rock. Thus he reported that there was some $7 million worth of ore in sight and that "the Chrysolite never looked so well, and beyond all question it has never been in so good condition for systematic, regular, economical, and large production." He further predicted that the mine would resume full dividends the following month.[35] Raymond made his report on May 20th. With such enthusiastic endorsement from so respected an authority, many investors wanted in and Chrysolite stocks began a steady rise.

Roberts and his partners, however, still faced a dilemma. The whole deception was likely to be exposed if the ore shipments from the mine were not increased to match Raymond's prediction; yet, if the last venire of ore was stripped from the walls of the drifts to provide those shipments, the scheme would be exposed just as surely when the barrenness of the rock behind was revealed. But the resourceful Keyes quickly hit upon a new ploy.

Just two days after Raymond's report, Keyes and Tabor took several local mining men on a personal tour of the mine to re-assure them of its richness. At the same time Keyes made pointed complaints about the slow progress in recent weeks which he blamed on a sudden laziness in his miners. His surveyor further testified to having seen the men "loafing, smoking, singing, telling stories and having a good time generally." Keyes then made a show of castigating his three shift bosses and issuing strict orders prohibiting smoking and talking during the shift. The surprised bosses denied the charges, stating firmly that "the men were doing their whole duty; . . . and that if Mr. Keyes insisted on the enforcement of his orders they would resign right there." Keyes accepted their resignations, posted the order, and left for Denver. He appointed George Daly of Bodie notoriety as acting manager to face the inevitable protest that would follow. The ensuing strike served Roberts well; it lasted for nearly a month, allowing him and his partners to unload much of their stock and at the same time skip another dividend without depressing stock prices or exposing the barrenness of the mine.[36]

But even though the Chrysolite management triggered and used the strike to facilitate their fraudulent stock promotion, it was the hardrock miners themselves who enlarged it to advance new causes of their own. At 7 A.M. on May 26th the night and morning shifts in the Chrysolite struck as expected for the re-instatement of the shift bosses and the rescinding of the prohibition on smoking and talking. But others soon joined them, swelling their number to six hundred, and one Michael Mooney began persuading the miners to make much broader demands for "an increase in our wages from $3 to $4 per day and a reduction of our working hours from ten to eight hours."[37]

Mooney was a miner in the Little Pittsburg and an unsuccessful Workingmen's Party candidate for alderman. He was branded a "Molly Maguire," but he denied this, although he proudly proclaimed, "I am a radical supporter of all that tends to ameliorate the life of the workingman." In the Chrysolite strike he saw an opportunity to take action on the broader issues of discontent which had been "laying dormant for months." Thus "when their contest came to a vigorous head this morning," he told a reporter, "I jumped into the breach and propose to make the best of it, be the result what it may." Mooney's rather precipitous entry into the strike aroused some suspicion and C. C. Davis, editor of the Leadville *Chronicle*, later suggested that Keyes and Daly had induced Mooney to inaugurate the strike. There is simply insufficient evidence to either substantiate or refute this charge. All that can be said for certain is that Mooney was the principal "agitator" of the broader strike and an able spokesman for its causes. He spoke for the western miners in their dissatisfaction with the mine managers who paid the same low wages to skilled and experienced miners as they did to "potato and grave diggers" from the east. And he spoke for the Knights of Labor who had secretly organized the Miners' Co-operative Union the preceding year in their efforts to establish eight-hour working shifts. The issues of the strike were thus broadened and it spread throughout the camp, helped along, of course, by George Daly.[38]

From the Chrysolite Mooney led the strikers to the adjacent Little Chief, also promoted by the "Roberts combination" and managed by George Daly. Just as they were about to go down and bring out the men, Daly arrived. He promptly discharged the crew and sent similar orders to the other mines under his and Keyes's management to do the same, discharging another two thousand miners. As a result of this action the Leadville papers for a time could not decide whether to call the dispute a strike or a lockout. The miners spent the remainder of the day tramping from one mine to another. By nightfall nearly every mine in Leadville was closed and more than six thousand miners were out on strike.[39]

The following morning about a third of the miners, led by Mooney and a brass band, paraded through the streets in a show of strength and determination. Afterwards they massed on Fairview Hill, just below the Chrysolite, and unanimously, resolved to demand a $4 minimum wage for an eight-hour day for all mine workers.[40]

Written notices of the demand were served on Daly and the other managers, who in turn telegraphed their directors in the east. The Chrysolite, Little Chief, and a few other companies were already working eight-hour shifts, but paid only $3 a day, whereas most of the companies worked ten-hour shifts and paid anywhere from $3 to $4 a day. A few of the smaller companies agreed to the miners' demands but most rejected them. George Roberts was doubtless delighted when he telegraphed Daly to "close mines indefinitely."[41]

From the start Mooney had cautioned the strikers to stay out of the saloons and refrain from any intemperate acts. He had also assured the superintendents that "no property would be touched," and that the miners were determined to prevent violence. But Daly seemed equally determined to prepare for, if not actually provoke, a violent showdown. The first night of the strike Daly hired gunmen to guard the Chrysolite, the Little Chief, and the Iron Silver, and he persuaded Sheriff Tucker to deputize them under the command of his Bodie lieutenants, the McDonald brothers. He also persuaded the sheriff to seize the rifles and revolvers in the local militia armory. These were later returned when the militiamen protested to the state adjutant general.[42]

At the same time Daly began fortifying the mines, making them "a veritable Gibraltar." The shaft houses were loopholed and lined with boiler plate; blockhouses were erected on the tramways and other commanding positions; supplies were stockpiled and preparations were made for a siege. When the miners learned of these actions, they sent a committee to the mines demanding removal of the gunmen, but they were rebuffed by Joe McDonald.[43]

The strikers then began to organize their own ranks. Since only a small fraction of the miners had belonged to the Miners' Co-operative Union, a Knights of Labor assembly basically opposed to strikes, the strikers decided to form a more militant union. On the morning of May 28th they again assembled on Fairview Hill to organize the Miners', Mechanics' and Laborers' Protective Association. Mooney read the constitution and bylaws from the pocketbook edition of the Virginia City Miners' Union and they were unanimously adopted with only "slight changes." Michael Mooney was elected president, John Crelley vice-president, and Andrew Lowman treasurer. Nearly all those present signed up. That afternoon a second meeting was held to "corral a few stray sheep not yet within the fold," but it was broken up early by a heavy snowfall. Despite the breadth of membership implied by its name, the new union still concerned itself principally with the interests of the miners.[44]

The Leadville press rather solidly aligned themselves with the mine management in condemning the strike and editorializing on the "absurdity of labor combinations." But in sympathy, a few local printers started a small, underground daily, *The Crisis*, on May 28th to champion the miners' cause. It was edited by George Clark and secretly printed in the *Democrat* office after that paper went to press. *The Crisis* succeeded too well, for within two weeks vigilantes ordered all those connected with it to leave the camp. The Leadville Typographical Union also pledged their support and assistance to the miners in their "battle against greed and selfishness," and, following the miners' example, the *Chronicle*'s newsboys struck for a higher percentage on their sales.[45]

Winfield Keyes returned to Leadville once the strike was well under way and in a show of determination he and Daly announced that they would resume work at their old rates on May 31st. Keyes, however, choosing not to personally reopen the mines, appointed Tim Foley as acting superintendent of the Chrysolite. Several other companies also agreed to resume and together they demanded protection from the county commissioners. The commissioners asked Sheriff Tucker to hire the local militia companies as guards at $4 a day plus rations. Although he had no authority to call up the militia, Tucker "directed" the Carbonate Rifles and the Highland Guards to go to the mines. But the Highlanders refused to "fight the miners." Tucker was furious; he discharged the company and asked the governor to have it disbanded. He then issued a proclamation to every able-bodied citizen of Lake County to stand in readiness to take arms in the event of violence.

Mooney criticized the employment of guards as "wholly unnecessary for the maintenance of law and order and . . . a reflection on the peaceable course pursued throughout by the strikers." Even superintendent Foley agreed, offering to give a bond of $500,000 that he could safely guard the mine himself without a revolver. But Keyes insisted on the guards, for the show of force was apparently intended more for prospective stock purchasers than for the striking miners. Chrysolite stock had been rising steadily throughout the first week of the strike and this show of determination on the part of the management sent it to a new high of $22¼ a share, nearly double the low it had reached just before the Raymond report.[46]

The miners, apparently fearing that this action might break the strike, offered to settle the dispute with a compromise of eight hours a day at 40¢ an hour. This slashed the miners' demanded wage increase from $1 to only 20¢ a day and was a major compromise. But Daly declared "emphatically that even had he the authority he would not consent to any compromise."[47]

Despite this rebuff Mooney cautioned the miners to remain cool, not to resort to force, and to rely solely on "moral suasion." With Mooney's assurances of no interference the Leadville papers jubilantly proclaimed the end of the strike, "BACKBONE BROKEN! THE SHIP OF COMMERCE HOISTS HER SAILS AGAIN!"[48]

"Moral suasion," however, carried the day as most of the miners refused to return to work. The Chrysolite and Little Chief each started up with sixty to seventy men, only a small fraction

of their full crews and nearly all unskilled laborers. These scabs
stayed in a boardinghouse within the barricades at the mines
where they were guarded by Daly's gunmen and Tucker's vol-
unteers, who were costing the county $500 a day. The guards,
in fact, outnumbered the scabs. A few other mines tried to start
up the following day, but were unable to get enough men. In
one mine fifty to sixty men had said they would come to work,
but only fifteen showed up. Of these only three had brought
their dinners and were willing to go down in the mine. The
superintendents claimed that the men had been frightened off
by threats and intimidation. But the only newsworthy incident
was a skirmish between a deputy and three roughs in a saloon
on one of the roads leading to the mines. Thus the once jubilant editors conceded, "the strike still continues to drag its
wearisome existence along with little probability of an early
dissolution."[49]

The miners on the other hand were greatly encouraged by the
managers' failure to break the strike. In closed meetings at the
city hall, they withdrew their previous offer and again demanded
$4 for an eight-hour shift for all underground men, but they
still made some concession, asking only $3.50 for surface work-
ers on the same shifts.[50]

A citizens' committee, however, called for much broader concessions, suggesting a "compromise," whereby the miners drop
all wage demands if the managers agreed to adopt eight-hour
shifts. This proposal, in fact, required no concession on the part
of Keyes and Daly, since they had already been working eight-
hour shifts. Thus they heartily endorsed this "compromise."
But the miners, needless to say, rejected it. And one observer
mused,

the greatest difficulty encountered by arbitrators in the settlement of
disputes of this nature is that the true character of the great mass of
laboring men is not understood, and efforts at settlement are made
invariably in favor of the wealthy, who are considered more shrewd
and cunning, consequently the bigger half must be given them. . . . If
the committee of "unselfish" citizens really take as much interest in
the welfare of the miners as it has expressed, why does it not try a
little of its persuasive powers on the mine owners, or at least seek an

Single jacking on the 500-foot level of the Tombstone Consolidated mine in Arizona, c. 1904. (Arizona Pioneers' Historical Society, Tucson)

Double jacking on the 1,500-foot level of the New Almaden mine in California, 1887. (Library, University of California, Los Angeles)

Machine drilling at the tunnel face 3,800 feet below ground in the Sierra Buttes mine
in California, 1888. (Bancroft Library, Berkeley)

Carman running out the ore from the stopes of the Wilson mine in Arizona, 1903. (Library, University of California, Los Angeles)

Scrip used to pay miners at the Ivanpah Consolidated mines on the Mojave Desert in 1881. (Mrs. Arthur N. Prater, Los Angeles)

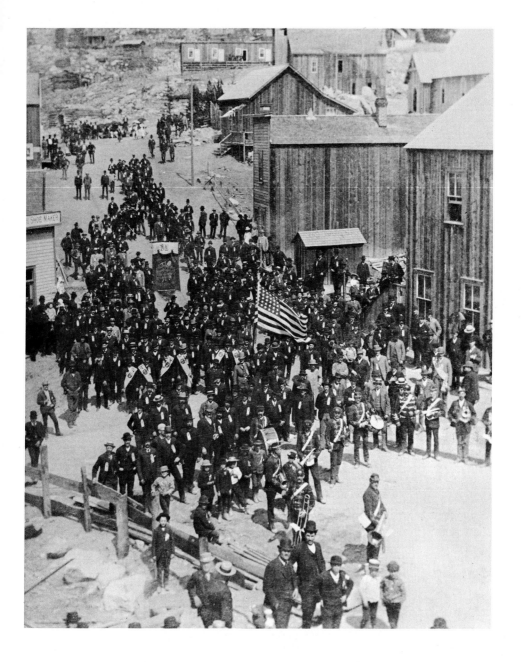

Miners' Union Day at Granite, Montana, 1889. (Montana Historical Society, Helena)

Stamp mill workers at Randsburg, California, c. 1900. (Title Insurance and Trust, Los Angeles)

A mining camp saloon at Randsburg, California, c. 1900. (Security Pacific Bank, Los Angeles)

Virginia City on the Comstock, 1890. (H. W. Fairbanks Collection, University of California, Los Angeles)

Imperial mine, where the Comstock miners won their first strike in 1864. Photograph by T. H. O'Sullivan, 1870. (Bancroft Library, Berkeley)

Gold Hill Miners' Union Hall, completed in 1870—the first miners' union hall in the West. (Library, University of California, Los Angeles)

William Woodburn

William Woodburn, president of the Miners' League of Storey County in 1864—the first militant miners' union in the West. (Library, University of California, Los Angeles)

Pocket editions of the miners' union constitutions and by-laws, which helped spread the mining labor movement throughout the West. (Bancroft Library, Berkeley)

Miners' Union Hall at Bodie, California, c. 1890. (Emil W. Billeb, San Francisco)

Leadville, Colorado, in the 1880s with the mines on Fryer and Carbonate hills in the background. (Library, University of California, Los Angeles)

Aaron C. Witter, first president of the Butte Workingmen's Union. (Montana Historical Society, Helena)

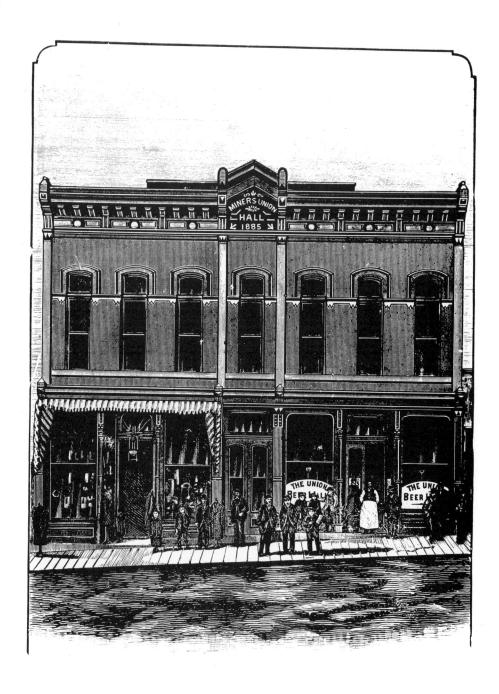

Butte Miners' Union Hall, built in 1885 and dynamited in 1914. (Library, University of California, Los Angeles)

Miners' Union Day parade, June 13, 1889, at Granite, Montana. (Montana Historical Society, Helena)

The mining camp of Gem in the Coeur d'Alene district, c. 1890. The miners' union hall
is the building with the balcony and flagpole in front. (Library, University of Idaho,
Boise)

Frisco mill at Gem before the battle between company guards and union miners on July 11, 1892. (Library, University of Idaho, Boise)

Ruins of the Frisco mill after a dynamite charge was sent down the penstock on the left. (Library, University of Idaho, Boise)

Federal troops marching into Wallace on July 14, 1892, after the battle at Gem. (Library, University of Idaho, Boise)

adjustment by which one side would not be asked to make all the concessions.[51]

As the strike dragged into its third week, the miners, beginning to weary of it, finally proposed once again to compromise at 40¢ an hour, or $3.20 a day on eight-hour shifts. The managers, however, again rejected this compromise and Keyes threatened not to reopen at all until September 15th.[52]

At this the editor of the *Democrat* became disenchanted with the managers and suggested that "the present half-hearted strike in Leadville is merely a stock jobbing operation, with Messrs. Roberts, Keyes and Daly in the wood-pile." C. C. Davis of the *Chronicle* also later charged Keyes, Daly, and a "coterie of operators" with starting the strike. But both considered it "a gigantic conspiracy to depress stocks," which it clearly was not.[53]

Chrysolite stock had, however, begun to decline after it became apparent that the widely heralded resumption of work had failed. By June 10th the stock had fallen off $5 a share from its high of $22¼ only a week before. It was obviously time for some dramatic new action. Thus in a whirlwind of excitement Keyes and Tabor prodded the Leadville merchants into organizing a vigilance committee to protect the mines; Keyes announced his departure for Denver to demand protection from the governor; and George Roberts declared he would ask the aid of the president of the United States. Two days later the vigilantes staged a show of force and the following day Governor Pitkin declared martial law. With this impetus Chrysolite stock climbed back to better than $20 a share before the miners finally capitulated, bursting the bubble.[54]

A number of citizens' committees had tried to negotiate a settlement of the strike, but Keyes's threat to close the mines for three months finally excited a number of merchants to take stronger, more direct action. They met with Keyes, Daly, and Tabor in the lieutenant governor's private rooms in his Opera House and formed a "Committee of Safety, whose objects shall be maintenance of order, the punishment of crime, and to take cognizance of all lawless acts that may transpire within our midst." A Council of Five, which apparently included Tabor,

Keyes, Daly, and Davis of the *Chronicle*, ran the committee. Publicly they styled themselves the Executive Committee of 100, although they earned the nickname of the "stranglers."[55]

On June 11th they proclaimed,

> Notice is hereby given by the undersigned, . . . that on the first step taken by any adherent of the Union, or other disturbers of the peace, to interfere with, intimidate or threaten any miner willing to work, THE UNDERSIGNED WILL SEE THAT SWIFT AND AMPLE PUNISHMENT IS METED OUT TO THE OFFENDER OR OFFENDERS.
>
> THE CITIZENS' EXECUTIVE COMMITTEE OF ONE HUNDRED.[56]

To give force to the proclamation the committee called for the organization of a "citizens' posse." That evening ten companies were formed and it was reported that the governor had dispatched the Pitkin Light Cavalry to bring them a thousand rifles. Sheriff Tucker in the meantime visited all the stores, confiscating guns. At Tabor's request Tucker also ordered the regular Wolf Tone Guards, made up principally of miners, to surrender their arms. They did so, then disbanded, protesting the insult.[57]

The vigilance committee effectively took over the town. They ordered all stores closed the following day and staged a Citizens' Law and Order Parade in an ill-starred effort to overawe the striking miners. The parade was to have begun before noon, but the rifles for the citizens' posse failed to arrive. Finally as the crowd grew impatient and rumors spread that "the momentous event had perished," the committee decided to march without the rifles and the posse scurried about for some sort of arms. The grand march got underway to a roll of drums at two in the afternoon. It was led by local Judge and mine owner A. V. Bohn and other committeemen on horseback, followed on foot by the posse of about six hundred "equipped in every imaginable way, some carrying pick handles, while others supported the weapon that had not seen sun light since its service in the revolution." Far from overaweing the strikers, the posse evoked only guffaws, hoots, and jeers as they passed.[58]

The Pitkin Cavalry finally arrived at four o'clock with two thousand cartridges but no rifles! They too paraded through the streets but with no better reception. Vigilante Bohn then took it upon himself to clear the street of strikers and onlookers. The crowd retreated to the sidewalk but refused to disperse. At this Bohn charged furiously into them, swinging right and left with his sword. Several in the crowd tried to drag him down from his horse before two policemen rushed in to arrest him and take him off to the police station.[59]

In the wake of the "Law and Order" debacle Sheriff Tucker telegraphed Governor Pitkin to declare martial law, charging that the Miners' Association had "made various threats against the lives and property of the citizens of this county, and have organized to carry out their unlawful purposes." Keyes, Tabor, and others also deluged the governor with similar pleas, and early Sunday morning, June 13th, he declared martial law in Lake County. Later, however, Pitkin regretted the decision, feeling that he had been deceived by "gross misrepresentations made to him by the mine managers and their friends."[60]

David J. Cook, western agent of the Pinkerton detective association, was appointed commander of the military forces in Lake County. He arrived at Leadville that Sunday evening to set up his headquarters in Tabor's private rooms at the Opera House. Immediately a flood of general orders began to pour forth from the headquarters, forbidding the carrying of weapons without a permit, closing all saloons and dance halls, outlawing the assembly of more than two people on the streets or in public halls, and clamping a 10 P.M. curfew on the town.[61]

The power of the vigilance committee was noticeably strengthened under martial law. Committeemen made up most of General Cook's staff. Their citizens' posse was temporarily mustered into the Colorado National Guard as its Third Infantry Regiment, and vigilante Bohn, who was out on $50 bail, was made colonel in command. Although Governor Pitkin had instructed Cook that "no committee or organization of either citizens or miners, can be permitted to interfere with the personal safety of the leaders of the other side," Cook made no effort to restrain the vigilance committee. They openly sent out notices to Mooney, Clark of the *Crisis*, and a good number of others reading,

Sir, You are hereby ordered to leave Leadville before sun-up tomorrow morning, to return no more. Disregard this notification at your peril.

By order
Committee of 100.

Mooney and the others realized that they had no choice but to leave.[62]

With equal efficiency the mine managers forced nearly all of the more militant strikers out of Leadville and crushed the strike. The managers blacklisted all who had "figured conspicuously in the strike," agreeing not to give them work in the mines, whereupon Cook began arresting them for vagrancy. Within a few days 250 men were arrested, while many times that number lined the roads out of Leadville.[63]

On June 16th the mine managers confidently reaffirmed "that their right to regulate the affairs of their mines is absolute, and that they will not submit to any dictation as to hours of labor or rate of wages from any person or association." Cook granted John Crelley permission to hold a final meeting of the Miners', Mechanics' and Laborers' Protective Association the following day. But although vigilante threats, blacklisting, and vagrancy arrests had decimated the association, the miners still refused to disband. Without debate, however, they agreed to call off the strike and to resume work at the old wages and hours. On behalf of the managers, Keyes and Daly agreed not to discriminate further against rehiring strikers. Daly, in fact, gave Crelley a job as shift boss in the Lowland Chief and Mooney also returned to work in the mines again. But when many of the other strikers applied for work, they found that most of the mines had already signed up full crews of greenhorns.[64]

Following the formal capitulation of the miners' association, Cook cut the Leadville troops to just two companies. They served until June 22d when the governor ended martial law. Hired gunmen were kept at the Chrysolite and a few other mines for some weeks, and the only fatality associated with the strike resulted from a shooting scrape between two of the gunmen at the Iron Silver mine.[65]

Immediately after the strike Keyes left for California on a

thirty-day leave of absence. During his absence the Chrysolite failed to resume its bonanza ore shipments and the realization slowly spread that the mine had been gutted. Most of the company's trustees soon resigned and it was revealed that the company was $400,000 in debt, much of which had been advanced to pay previous dividends. On his return Keyes also resigned. At the same time Roberts and his "coterie" dumped the last of their stock on the market. The stockholders panicked and the stock collapsed. Chrysolite became the most active and demoralized stock on the New York exchange. Weekly trading reached 116,000 shares, over half the total stock issued, as everyone tried to get out from under it before it bottomed out at $4 a share.[66]

Although the exact figures are not available, Roberts, Tabor, and their confederates apparently unloaded about 150,000 shares on the public, taking in nearly $3 million from the fraudulent promotion and manipulation of the Chrysolite stock. Roughly $1 million of this was gleaned during the strike, which helped sustain the stock prices so they could unload at high figures. This booty was well in excess of the $1.1 million in profits actually earned from working the mine. Daly was reported to have made $60,000 in Leadville; how much more Keyes made is unknown.[67]

Raymond's *Engineering and Mining Journal* denounced the "Roberts Combination" and their "disgraceful and criminal proceedings," warning "it makes no difference how intrinsically valuable a mine may be, when it is managed by tricksters and stock-gamblers it is always a bad investment." Still Raymond was slow to appreciate how badly he had been deceived in his valuation of the mine, and hoping to prove its worth, he became one of the new trustees of the company. But even with the discovery of additional ore bodies the subsequent production of the Chrysolite reached only a small fraction of Raymond's estimated $7 million and he finally realized that he too had been taken.[68]

Confidence in other "Roberts stocks" was broken by the Chrysolite swindle. Little Chief stock which had been running at close to $10 a share collapsed to 60¢ that summer. Soon after the collapse George Daly left Leadville to manage other Roberts promotions. He was killed by Apaches the following summer in New Mexico at the Sierra mines which Roberts was floating in Philadelphia. A decade later Daly's henchman, Joe McDonald, turned up behind the barricades against striking miners again in the Coeur d'Alene district of northern Idaho.[69]

Despite the fact that the strike was instigated and used by the Chrysolite management only to maintain their stock at a false level, the failure of the strike was a real defeat for the mining labor movement in Leadville. The Miners', Mechanics' and Laborers' Protective Association did not disband immediately, but it was so weakened that it soon disintegrated. Even when the Leadville miners did reorganize nearly five years later, they formed only another conservative assembly of the Knights of Labor which acquiesced to further wage cuts. Militant unionism was dead in Leadville for over a decade.

But devastating as the Leadville defeat was, it was the only major battle lost out of many won as the mining labor movement spread throughout the west in the late 1870s and early 1880s. And the militant miners driven from Leadville joined in the labor battles elsewhere with new determination.

The Four Dollar Fight

The middle 1880s were marked by a fight against the $4-a-day wage in the hardrock camps of the intermountain West. The miners lost, and the long and bitter fight dealt a crippling blow to the western mining labor movement, as many of the unions themselves fell with the $4 wage. The fight was triggered by the decline of silver and lead prices, which forced the closure of numerous mines and retrenchment in many others. Retrenchment for most mine owners meant a wage cut for the miners. The miners' unions resisted and the struggle frequently flared into violence as the issue became the very survival of the unions themselves. A few of the unions were strong enough and prudent enough to gain a compromise and they survived, but many failed. Thus as the growth of the western miners' unions had followed the expansion of the silver mining industry, so too did the temporary decline of the unions follow the decline of metal prices and the retrenchment they forced upon the industry.

The first rumblings of the fight came well in advance of the precipitous metal price declines and were the reverberations of the falling ore values on the Comstock lode following the exhaustion of the bonanza ores. What started as a call for a general wage cut on the Comstock changed by early 1881 into a clamor for lower wages only in the upper levels of the mines to allow the opening of large bodies of lower-grade ore. The Comstock miners' unions rejected a wage cut, but they prudently suggested an acceptable compromise in allowing miners in the upper levels simply to work longer shifts for their $4. Work commenced on this basis in March of 1881, ending the agitation for a wage cut on the Comstock for more than a decade.[1]

Elsewhere in the West, mine operators also began calling for a general attack on the $4 wage. One enthusiast became nearly ecstatic as he wrote to the New York *Mining Record*,

The next blow must be against the Miners' Union and extravagant wages. . . . With wages at three dollars per day, we could get men quite as competent, and mines which are lying idle to-day would be worked, and mines which are a drain upon the pockets of men who have invested their hard earned dollars in them would be paying good interest . . . and many an honest fellow who is struggling for bread in the crowded East could find a home and work and wages to support it in our silver lined mountains. To be sure, the Cornish man would have less money to waste on beer and harlots, but the hill sides would be covered with little homes and little smiling children, and the ample wages of the father would keep want from the door and nature would yield her treasures in greater profusion, and year by year millions would grow to billions of glittering coin to be thrown into the coffers of the world.[2]

Although many western mine owners had looked forward to the possibility of a wage cut, they had waited to see the outcome of that issue on the Comstock before attempting any action on their own. One freshman mine owner, however, fired up by all the rhetoric, just couldn't wait. William S. Clark, an Amherst College chemistry professor, had gotten the mining bug in 1880. In partnership with a former newspaper reporter, John R. Bothwell, he talked some of his fellow faculty and a number of wealthy widows and farmers into investing in western mines. They put $250,000 of their savings into the Starr-Grove mine at Lewis in central Nevada and the Stormont mine at Silver Reef in southwestern Utah. Clark became president of both companies and Bothwell secretary. They fired the former superintendents, appointing Bothwell's brother George to run the Starr-Grove and an acquaintance, Colonel W. I. Allen, to the Stormont.[3]

Touring his mines in January of 1881, Clark happily telegraphed east that the Stormont was producing between $40,000 and $50,000 a month at a cost of only $6,500, while in the Starr-Grove "$1,000,000 worth of ore could be taken out at an expense of only $50,000"—profits of seven- to twenty-fold! Yet even such phenomenal profits were not big enough, for he gave orders to cut wages 50¢ a day on February 1st.[4]

In both camps the miners struck against the wage cut, but the two strikes were almost mirror images of each other. Before the strike the Lewis miners were not organized and received only $3.50 a day, but, winning the strike, they emerged with a strong union and a $4 minimum wage; whereas the Silver Reef miners before the strike had a well-established union and a $4-a-day wage, but, losing the strike, their union was badly weakened and their pay was cut to $3.50. The difference between success and failure in the two camps resulted from many factors, but the solidarity of the Nevada unions, even to the point of direct intervention, and the lack of community support in the Mormon settlements seem to have been the crucial factors.

Lewis was a small but growing camp in early 1881 with only about a hundred working miners. Nearly half of them were employed in the Starr-Grove, which was the principal mine in the district. When Bothwell announced the wage cut from $3.50 to $3 a day, the miners first struck for the reestablishment of the old rate. But within a week Martin Igo, president of the Austin Miners' Union, had come to Lewis and organized the strikers into a union to demand a $4-a-day minimum wage. With this demand Lewis's second largest mine, the Betty O'Neal, also closed down, but all of the smaller mines agreed to the new rate.[5]

After a stalemate of over a month Bothwell offered to compromise by resuming work at the old rate. The miners' union voted to reject the offer, but the men were divided and a number withdrew from the union. The dissidents joined with non-union miners to accept Bothwell's proposition. Violence erupted on March 9th when three men were wounded in a shooting scrape after a union man called one of the deserters a traitor. That same day thirty-five scabs, accompanied by the constable, attempted to resume work in the Starr-Grove, but were turned back at gunpoint by the strikers.[6]

Immediately after the shootings the constable telegraphed the deputy sheriff at nearby Battle Mountain for "help and arms." At the same time the Lewis union's president, William Tohill, telegraphed the Battle Mountain and Austin miners' unions for aid. The deputy sheriff's posse arrived by a special train the following evening, found all quiet, and left the next morning. In the meantime the Austin Miners' Union had voted to go en masse to "bring about an adjustment of the difficulties." Thus some 250 Austin miners, armed with everything from pistols to shotguns, boarded another special train headed for the scene.[7]

The deputy sheriff's posse left Lewis just before the Austin miners arrived; so did superintendent Bothwell. Fearing "the little war cloud on the southern horizon," he gathered up his family and worldly goods, and headed for San Francisco. But before he left, he capitulated to the union's demand, ordering his foreman to resume work that same morning at $4 a day with "none but union men." Thus the Austin miners forced capitulation just by the threat of intervention. After a wild night celebrating, they headed home to be greeted "like a regiment returning from a victorious campaign with all the pomp and circumstance of glorious war." Lewis and its union grew rapidly after the strike and by May there were three hundred men working in the district.[8]

Silver Reef on the other hand was a larger, better-established camp than Lewis in early 1881. Its mines had produced around $3.8 million in silver during the previous four years, better than half of it by the Stormont which had paid out $155,000 in dividends; and its miners had organized a union in February 1880, with three hundred members and a $4 minimum wage. Thus on February 1, 1881, when the new superintendent, Colonel Allen, told the miners going on the morning shift, "You can't go down unless you go for $3.50," they struck. Superintendent Ogden of the Barbee and Walker mine, sympathizing with the cut, discharged half his men and vowed not to hire new men for more than $3.50 a day, although he still paid $4 a day to the old hands. But the other superintendents, who employed about half the men in the camp, opposed the cut and continued at the old rate.[9]

The strike dragged along without incident until late February when Ogden announced that he would shut down the Barbee and Walker completely until the miners agreed to the cut. Fearing an enlargement of the strike, the miners decided to try intimidating Ogden by making an example of Allen. On February 27th they marched on the Stormont, gave Allen just five minutes to grab his belongings, and escorted him out of town. They then warned Ogden that if he closed down they would "serve him the same." Ogden got the message, deciding it was "good policy to keep quietly at work for the present."[10]

News of the incident made Salt Lake *Tribune* editor C. C. Goodwin's "blood boil with shame and indignation." The union's "leaders are not Americans," he charged. "They must be men from a foreign country and the blood of five hundred years of serfdom must taint their veins else they would not so transgress every principle on which this Republic is founded." He further accused them of Molly Maguireism and reminded them that "the Molly Maguire leaders, to the last man, were hung like dogs."[11]

The Silver Reef miners did not meet the same fate. But also unlike their comrades at Lewis, they did not go unpunished. Allen got indictments against a number of the miners on charges of riot from the federal grand jury at Beaver. A U.S. Marshal was sent to arrest them. He recruited a posse of twenty-seven from the Mormon farming community of St. George and staged a surprise raid on Silver Reef on March 16th. Riding into town under cover of a light snowstorm, they arrested the miners on their jobs at the Barbee and Walker. Twenty-two union men, including president A. H. Lewis, were placed in irons and all the saloons were closed as a precaution against further trouble. The prisoners were tried at Beaver and thirteen were found guilty and fined a total of $1,225 to foot the bill for the marshal's and the court's costs.[12]

The arrests seriously weakened the union. Although they received financial aid from the Ruby Hill and other miners' unions and held nightly meetings to try to maintain morale and solidarity, an observer noted "the cheering grows perceptibly fainter." Within a week Allen returned from exile, spurring Ogden to finally shut down the Barbee and Walker until the wage dispute was settled. Fear of violence prompted the formation of a secret Silver Reef Safety Committee which issued an impassioned call to arms, but most "laughed the matter off."[13]

The strike dragged on for another month. It was a war of attrition. A few more miners left every day; a couple of merchants who had extended credit to the miners went bankrupt; and Wells Fargo and Company and a local merchant attached the Barbee and Walker. But the miners who stayed were hit the hardest and they finally gave in. On April 24th the union voted to "accept the Stormont proposition ... and hereafter work for $3.50 instead of $4 per day."[14]

The editor of the Silver Reef *Miner* concluded in disgust "there are times when it is but fair to attribute a 'method in madness,' but this whole business has demonstrated a lack of wisdom verging on the ridiculous. ... The mere reduction of half a dollar a day will avail the company little, and certainly not place it on a dividend-paying basis, if other and more needed reforms are not brought to bear on the old system of management."[15]

The Stormont resumed work but its management was not noticeably improved. Within a year the stockholders of the Stormont, the Starr-Grove, and five other mines, which Clark and Bothwell had dabbled in, all brought suit against them for the misappropriation of $325,000. That was the end of Professor Clark's adventure in mining.[16]

An underlying cause of the union's defeat at Silver Reef was their proximity to the Mormon farming communities. Farmhands, willing to work seasonally in the mines for cheap wages just to pick up a little cash, broke the union's solidarity and the religious antagonism between the Mormons and the "gentile" miners deprived the union of broader community support. Thus the loss of the Silver Reef strike also broke the $4 wage in Park City and in the other camps around the Mormon settlements in Utah. But the miners' successes in the Lewis strike and the Comstock dispute deterred, for a time at least, further assaults on the $4 wage elsewhere.

No union action, however, could halt the decline of metal prices. Silver, which had held at close to $1.14 an ounce following the passage of the Bland-Allison silver purchase act of 1878, took a sharp drop to $1.08 in the fall of 1882. At the same time lead began a disastrous decline that took it from $5\frac{1}{8}$¢ to $3\frac{2}{3}$¢ a pound by the end on 1883. Silver prices made only a slight recovery before they began a long crippling decline in 1884. These declines forced drastic economies in the mining industry and gave strong impetus to a new assault on the $4 minimum wage.[17]

The first blows were again struck through the press. Spokesmen for the mine owners assailed not only the $4 "arbitrary standard" but the miners' unions themselves. "This pernicious association," the editor of the San Francisco *Exchange* charged, "has done more to destroy our mining industry than anything else, and to it is in a great measure due the decadence of our mining interests, for it cannot be denied that whenever this tyrannical society is in power, the miners are not prosperous or the mines profitable." But former miner turned editor, William Penrose, blasted back,

mine owners, speculators and others who have grown fat from the dishonest management of our mines are very willing to let the blame for the decline of a camp rest upon the Miners' Union. If the management has stolen the proceeds of the mine, if the ore has given out, if water has come in in such quantities that it cannot be handled, if the company has been robbed until it cannot afford to put up new machinery when required, if anything has happened to lower the stock or decrease the bullion output, the fault is laid at the door of the Miners' Union.[18]

In the same spirit the *Mining and Scientific Press* sought to blunt the assault on miners' wages by suggesting that the management first put its own house in order.

In looking over the reports of companies, it usually strikes us that such items as legal expenses, sundries, salaries, etc., go to swell the expenses more than they should. If expenses are to be reduced—and they should be—the top-heavy system of mining too much in vogue will be best remedied by taking off the weight from above. A trimming of salaries and rigid inspection of general expenses will serve a better purpose than cutting down the wages of the men who do the real work.[19]

But the decision to try to cut miners' wages was in the hands of the owners and directors, and a few at a time as the circumstances seemed to demand, or as the time seemed ripe, they made that choice. Some of the companies, either through overlapping directorates or through mine owners' associations, acted in concert in cutting wages. By 1884 a Nevada editor observed "there seems to be a move throughout the whole of the inter-mountain country tending to the reduction of the price paid labor of all kinds, and it is gradually being accomplished." But as Nevada was the fortress of the $4 wage its camps were the last to feel the assault.[20]

The first attacks came on the perimeters on the intermountain country, in the San Juan on the east, at Tombstone on the south, and at Wood River on the north—and one after another they grew in intensity. In August of 1883 the mine owners in southwestern Colorado formed the San Juan Miners' Association "to promote the varied mining interests of the San Juan in every possible way." One of the first ways they had in mind was a wage reduction. On November 1st, just before the onset of winter, the mine owners at Telluride jointly cut miners' wages from $4 to $3.50 a day. The miners' union opposed the cut and struck, closing every shipping mine in the district. After a few weeks some of the mine owners proposed to compromise at $3.75 a day. The union agreed, provided the mines would reinstate the strikers. But the owners rejected this condition and put scabs to work at $3.50. The strikers held out for over a month before winter finally forced them to capitulate. Two months later the $4 wage fell at nearby Silverton. The editors of the *Engineering and Mining Journal* coyly saw these "local contests" as the "forerunners of a more general discussion of the relations of employés to employers." But they also predicted, "knowing the temper of their men, the majority of the mine owners of the West will be forced to bring their miners to a realizing sense of the gravity of the situation by closing down, whenever they can do so."[21]

Along the southern border at Tombstone, Arizona, the miners staged a much more protracted resistance in the spring and summer of 1884. Tombstone had been a $4 camp since its discovery in 1877 and its principal mines, the Contention, Grand Central, and Toughnut, had produced over $10 million in silver and paid over $4 million in dividends. Nonetheless the Tombstone miners had not succeeded in organizing a union because of the determined opposition of the mine owners who vowed that "a Union will not be permitted" and discharged men even "for favoring the organization of one." But when the Grand Central, Toughnut, and Contention posted notices that starting May 1st they were cutting wages down to $3 a day, the miners felt they had nothing more to lose. On April 29th they organized the Tombstone Miners' Union, resolving to oppose the reduction. More than three hundred miners, roughly three-fourths of those in the camp, joined the union. They adopted the constitution and bylaws of the Gold Hill Miners' Union and elected S. D. Stevens president and S. R. O'Keefe recording secretary.[22]

The mining companies, however, were equally determined to enforce the reduction. Superintendent E. B. Gage closed the Grand Central mine the following day and the Toughnut and Contention closed a few days later. All announced that they would not resume for more than $3 a day. Gage claimed that the only ore left above the water level was running less than $20 a ton and would no longer pay at $4 wages. Although richer ore was known to exist below the water level the companies were not yet willing to invest in adequate pumps to control it.[23]

Most of Tombstone's citizens sided with the miners' union at first. On May 5th after the strikers had staged a parade through town, the merchants treated them to three barrels of beer and a thousand cigars. That evening a mass meeting was called by sympathetic citizens to raise a subscription for a strike fund. Despite their aid, however, Stevens conceded, "I hope that the matter will be settled peaceably and speedily, as I am not in a position to sustain a strike." Most of the strikers were living on very meager savings and many lost even that on May 10th when the bank of Hudson & Co. closed its doors owing Tombstone depositors $130,000. As an angry crowd demanded that the bank open its doors, the grand jury ordered Sheriff J. L. Ward to "permit no one to enter or remove anything." Ward, however, secretly withdrew nearly $7,000 for himself and his friends, which caused a new sensation when it was finally discovered.[24]

The bank closure temporarily paralyzed Tombstone business and a citizens' committee was appointed to try to negotiate a speedy settlement of the strike. They proposed that the union accept the pay cut if the companies would promise to raise wages again once pumps were put in and work was commenced below the water level. The miners apparently agreed. But the companies only repeated their determination not to resume at more than $3 a day and the directors of the Grand Central added, "This is final!"[25]

Thus the Tombstone Miners' Union prepared for a long siege. The Virginia City, Gold Hill, and Bodie unions all contributed money to sustain the strike fund, as did the hundred or so working miners in the Tombstone unions who were still making $4 a day in the smaller mines. In addition organizers went to the neighboring camps of Bisbee and Globe to help form miners' unions there which aided the Tombstone strikers. But many of the miners decided to get out while they still had traveling money.[26]

In an effort to put some pressure on the Grand Central company, the miners' union also tried boycotting N. K. Fairbanks & Co., the Chicago lard packers, and the mine's principal stockholder. They solemnly resolved to "abstain from the use of all goods of every description manufactured by N. K. Fairbanks & Co., their heirs and assigns, until such time as they resume work in the mine at the former rate of wages, even though it be forever." But they failed to trim Fairbanks's fat sales.[27]

As the strike neared the end of its second month, Tombstone business was severely depressed. The Tombstone *Epitaph's* editor, who had previously taken no stand on the strike, finally tried to convince the miners to capitulate. He wrote, "These words are not written in the interest of the mine owners and superintendents, who have left us to our fate, but for the benefit of the miners and business men of this camp. There is no denying the fact that there is not one in ten of us who can last a month longer. . . . The day has passed when even a compromise at $3.50 could be effected; now the only salvation for Tombstone camp seems to be to accept the situation, hard as it is to bear."[28]

Most of the union men were still as determined as the owners, however, so the deadlock continued for another month. But as the days dragged on and the strike became the longest yet in the West with no end in sight, a growing number of miners became willing to settle for the cut. Thus on July 25th the Grand Central company finally started up with fifty men at $3 a day in the Head Center mine under superintendent C. M. Batterman. The union weakened by defections made no immediate response but within a few days union president Stevens called a citizens' meeting in Schieffelin Hall—now a museum. Announcing that he had "lost control" of the union, he called upon the citizens to appoint a committee to ask Batterman to stop work in the mine. "Unless this was done," Stevens warned, "he would not be answerable for the consequences." Although most of the citizens had supported the miners at first, by now they too were desperate for an end to the strike, and the resumption of work, even if it meant defeat for the union, would bring an end. Many also resented Stevens's threat, although he denied that was his intent. Thus they refused to intercede and the meeting adjourned.[29]

Rumors quickly spread that the union planned to hoist the scabs out of the mine. Sheriff Ward posted deputies at the Head Center and barricaded the works. But after considerable debate, the union sent only a committee of three to try to negotiate with Batterman. Confident of success, he refused to pay more than $3, or even to hire union men at that rate, unless "they severed their connection with the union or the union disbanded."[30]

The Grand Central company, advertising throughout the country for "good miners" at $3 a day, got scabs from as far away as Pittsburg. On August 4th they put sixty men to work in the Grand Central mine. With this the editor of the *Epitaph* concluded "the strike is at an end." But he commiserated with the miners, and praised the "law-abiding spirit shown by the union men during the three month's lockout. The course pursued by the union in the main has been temperate and manly, and the harsh words, if any have been spoke, should be forgotten."[31]

The union was, in fact, broken, but those who still clung to it, embittered by the struggle, refused to concede defeat and soon violence erupted. On August 8th one of the strikers made "very boistrous demonstrations" on the street, and when he resisted arrest, an officer dealt him a crushing blow with a revolver butt, fatally fracturing his skull. That night another union man got into a shooting scrape with a scab. As a threatening crowd gathered, the sheriff put the scab in jail for his own safety. The crowd refused to disperse, but drifted off in smaller groups. "Very ominous" rumors spread through the town and a general anxiety and uneasiness prevailed. Shortly after 3 o'clock in the morning, when the night shift left the Grand Central, the guards noticed several dozen men wending their way up the ore road

toward the hoisting works. The guards retreated to the works
and prepared for a fight. When a union man came to the door,
asking the foreman, C. W. Leach "if he intended to continue
work at $3 a day," Leach took him prisoner and the shooting
started. In the confusion the prisoner escaped and the attackers
retreated. No one was injured.[32]

The editor of the *Epitaph* bitterly charged, "The cowardly
attack made this morning on the Grand Central works has removed every particle of sympathy which this community here-
tofore held for the Miners' Union, or rather the handful of
men who still claim allegiance to that organization." That afternoon several of the superintendents, headed by Batterman,
served an ultimatum: "To whom it may concern: Unless the
Miners' Union of Tombstone formally disband before 3 P.M.,
Sunday, August 10th inst., and deliver over to the sheriff the
ring leaders of the cowardly assassins who made an armed attack on the Grand Central Hoisting Works last night, no man
whose name is now on the roll of said Union can ever have any
work at any price." But the last holdouts in the union still refused to disband.[33]

Backed by the superintendents, the sheriff used the attack as
cause to appeal to the acting governor H. M. Van Arman for assistance. Two companies of federal troops were ordered to Tombstone from Fort Huachuca. They arrived on the morning of
August 12th and went into camp at the mines on Contention Hill
"to remain until all fear of trouble is over."[34]

After two more weeks even the staunchest and most loyal
union men realized that further resistance was futile and on
August 25th the union finally disbanded. About $1,000 still
remaining in the treasury was donated to set up a fund for sick
and disabled miners. Thus after a struggle of nearly four months,
the miners were finally forced to capitulate to the $3 wage. As
the strike had dragged on with ever increasing financial loss to
community, their support of the miners slowly eroded, and with
the desperate attack on the Grand Central the last vestiges of
that support were swept away. At the same time the union
solidarity was undercut by those $4 men who chose simply to
move on rather than staying to fight.[35]

Reflecting local opinion the editor of the *Arizona Star* pro-
nounced the strike "a deplorable failure," concluding "no one
has been benefited and the miners, the mine owners, the merchants and the entire county of Cochise has suffered great injury, and Arizona in general has felt the blow given to the mining industry." But the *Engineering and Mining Journal*, sharing
the viewpoint of the mine owners, saw the defeat of the $4 wage
as a major victory, suggesting "Tombstone has taken the lead
that more than one mining camp will be forced to follow." The
other Arizona camps did, in fact, soon follow. Wages were cut
at Bisbee in October of 1884 and at Globe in January of 1885.
Concluding from the Tombstone struggle that opposition was
futile, the Bisbee and Globe miners' unions quietly acceded to
the cuts. By the end of February 1885 miners' wages had been
reduced throughout the territory to $3 a day. Miners' wages in
Utah were also cut from $3.50 to $3 at the end of 1884.[36]

An exodus of skilled miners followed the wage cuts. Among
those who left Tombstone soon after the strike began was Barney McDevitt, night foreman in the Head Center. At thirty-
eight he had worked half his life in the $4 camps of Nevada and
California, and when wages were cut in Tombstone he simply
moved on to hunt good wages elsewhere. Putting Tombstone
far behind him, he went north to Wood River, Idaho, where the
$4 wage still seemed secure. Ironically he would soon become
the leading figure in the last major fight to uphold the $4 wage.[37]

The rush to Wood River had begun in 1881. By 1885 nearly
six thousand people had settled in the camps along the river,
and the mines had shipped over $8 million in silver-lead ore
and were developing steadily. Hailey, the Alturas County seat,
Ketchum ten miles north, and Bellevue five miles south, all
located on a newly completed branch of the Oregon Short Line,
rivaled one another to be the principal business and shipping
center of the district. Most of the working miners, however,
lived either at Bullion in the hills seven miles west of Hailey or
at Broadford just across the river from Bellevue. The remainder
were scattered around Ketchum and the smaller camps of
Vienna and Atlanta some forty miles to the northwest.[38]

A miners' union had been organized at Bullion with the first rush in the summer of 1881 and one at Broadford the following year. The two unions had a combined membership of roughly 450. They also had the enthusiastic support of T. E. Picotte, the editor of the Hailey *Wood River Times* and a former president of the Mechanics' Union of Lyon County at Silver City on the lower end of the Comstock. Picotte had alerted the unions in June 1884 to the impending threat to wages posed by the discharge of hundreds of railway laborers after completion of the Oregon Short Line. Most of the superintendents, however, wanted only experienced miners, but one, Cecil B. Palmer, just arrived from England as superintendent of the British-owned Minnie Moore mine at Broadford, was eager to take advantage of the cheaper labor. On July 20th he cut wages to $3.50 and the Broadford Miners' Union promptly struck, closing down the mine. The other superintendents refused to join in the cut and after only five days Palmer restored the $4 wage. Picotte congratulated both the miners' union on their first victory and Palmer on "his manly course" in conceding defeat. Thus the $4 wage seemed secure on Wood River when Barney McDevitt and others flocked in.[39]

But Palmer did not really accept the defeat and six months later in the dead of winter he determined to try cutting wages again. But this time he followed a more devious strategy and he convinced A. J. Lusk, superintendent of the Queen of the Hills mine, to join him. In mid-January of 1885 Palmer laid off all his wage men, keeping only tributers in the stopes and contractors on development work. Then he and Lusk announced that after the 20th they would pay no more than $3.50 a day. McDevitt, working in the Queen of the Hills, was one of those whose wages were to be cut, but this time he was not willing just to move on. He had, in fact, been elected president of the Broadford Miners' Union and with the others, confident from their earlier success, he was now determined to stay and fight. The miners rejected the cut and called a strike. The Queen of the Hills, which employed only wage men, was closed down, but the unwary union men allowed Palmer to continue work in the Minnie Moore with the contractors and tributers not directly affected by the wage cut.[40]

The miners soon realized their mistake, however, when Lusk
signed up a dozen new contractors to resume work in the Queen.
As soon as he learned of the new contracts, McDevitt called an
emergency meeting of the Broadford union. He denounced the
letting of new work on contracts as an "opening wedge" to
break the strike and the union. Even the union miners, who had
been working on contract in the Minnie Moore and making $4
or better a day, agreed to quit for the duration of the strike. Thus
the union announced that it was "against their wishes" for any
work at all to be done in the mines until the trouble was settled.
All the contractors and tributers heeded the request and both
mines were closed. The Bullion union solidly supported the
action of their Broadford comrades, announcing "the Bullion
and Broadford Miners' Union WAS AND IS ONE, and as such we
sustain one common interest together for the good of all." But
community support for the action was much less solid. Although
most of the community had opposed the wage cut, many failed
to appreciate the subtler threat posed by working the mines on
contract and they considered that the miners had gone too far
in opposing it.[41]

In Bellevue, Lusk soon rounded up several new contractors,
who were willing to oppose the union. On February 9th three
of them quietly started work in the morning. But when they
attempted to return to the mine after lunch, they were stopped
by McDevitt and several dozen others lining the snowbank
above the trail. That evening the other contractors came over
to Broadford. They were also met by McDevitt, Charles O'Brien,
the past president of the union, and several other union men.
McDevitt asked if they too had "come to go to work on that
contract?" When they replied that they had, O'Brien said they
were "queer men to come down there to work against them."
He explained that the contract was only a "blind," so the company could break up the union, and asked them to quit. When
they refused McDevitt told them, "We'll give you just five
minutes to leave town." A large crowd quickly surrounded them
and marched them down the street toward Bellevue. Just before
they reached the bridge over Wood River, McDevitt suggested,
"We'd ought to make you get down on your knees and take an

oath." The crowd hollered back, "Make them get down!" Drawing his pistol, McDevitt commanded, "Down on your knees, and hold up your hands! You solemnly swear that you will never cross this river again to undermine this Miners' Union." They swore and were sent on their way, as one miner yelled after them, "Take that contract back to Captain Lusk, and tell him to take it to the privy with him when he goes."[42]

Learning of the outrage, Lusk demanded protection from the sheriff, Charles Furey. He dispatched deputies to both mines the following morning, but work was not resumed. A few days later Furey arrested McDevitt and eleven others on charges of "unlawful and felonious conspiracy." After a lengthy preliminary hearing before the probate judge, McDevitt, O'Brien, and five others were held over for the grand jury. Picotte helped put up bail for McDevitt and others, but he wrote,

HALT!

That is what Judge Street's action in holding the Miners' Union boys over to appear before the Grand Jury says to them: Halt! you have gone to far. You have overstepped the bounds beyond which no law-abiding citizen can go, and must stop. . . . The "difficulty," "trouble," "strike," call it what you will, should be submitted to arbitration. Let the coolest heads on both sides step to the front and see what can be done.[43]

Picotte's call for arbitration went unheeded. On February 24th, the morning after the hearings ended, Lusk and Palmer ordered their contractors back to work. Palmer personally tried to escort them up to the mines, but they were met by a crowd of some thirty miners, who told him that he could go to the mine, but "not one man could go with him." Palmer returned to Hailey to swear out warrants for their arrest. That evening sheriff Furey arrested sixteen for conspiracy. At a mass meeting the following evening the union men promised to "keep within the law and voluntarily and instantly squelch any disturbance," but they still swore to hold out for $4 a day.[44]

Despite these assurances, Lusk and Palmer demanded protection from the sheriff before they attempted to resume work again. Furey complied, raising a posse of Bellevue and Hailey men to escort seven contractors back to the mines on February

28th. More than a hundred miners lined the street in Broadford, laughing, hooting, and yelling as they passed, but no one tried to stop them.[45]

The union men were still confident that with "moral suasion" they could eventually "talk them out of the notion of working." They were also well prepared for a long siege. The Broadford strikers received financial aid from the working miners in their own union, as well as the Bullion union, and Hailey and Bullion merchants donated nearly $1,000 to the strike fund. In addition the Virginia City and Bodie miners' unions gave them financial assistance, and secured Nevada congressman and pioneer labor organizer William Woodburn and Bodie miner-lawyer Pat Reddy to help defend the union men charged with conspiracy.[46]

But the "wedge" had been driven in the strike. Within a few days a dozen more "sheepherders and cowboys" went to work, although they were guarded full time by three dozen deputies at a cost of $170 a day to the county. "Moral suasion" had little effect on these scabs and by mid-March the Minnie Moore had seventy-five men and Queen of the Hills had sixty. At the same time a number of the strikers, tiring of the fight, had left Broadford to seek $4 wages elsewhere. Even past president Charles O'Brien joined the exodus to take a job at Bullion. Some dissension also grew within the union and a few who were perhaps too eager to settle for $3.50 were expelled. Several of the strikers got into fights with the scabs, leading to more arrests. Rumors spread in Bellevue that the miners were going to dynamite the town and a Committee of Safety, dubbed the "stranglers," hastily formed to patrol at night. Counter rumors soon spread that the stranglers intended to hang the union leaders.[47]

The miners made one last effort to drive out the scabs in a massive demonstration on St. Patrick's Day. A call was issued throughout Wood River for all sympathetic to the cause to join in. A special committee went to Ketchum to organize a union there to help. The day before the big event the "boys" began arriving from the surrounding camps. As one remarked, "We have come to settle this thing if it takes all summer, but we are not going to make any outbreak, or commit any crime." Still Palmer and Lusk, fearful that the miners intended to "capture the mines," persuaded the sheriff to send over more rifles and more deputies to man the bulkheads guarding every approach to the mines.[48]

Most of the outside miners, numbering over a hundred from Bullion and fifty from Ketchum, arrived shortly before noon on St. Patrick's Day. Joined by the Broadford miners, they paraded through town, then massed at the union hall to appoint a committee, headed by McDevitt, to confer with the superintendents. At first Palmer refused to meet with the miners, claiming it was "beneath his dignity to discuss matters of business with such open and notorious violators of the law." But his legal counsel finally persuaded him to submit to the indignity, so the miners could not claim that they were willing to settle but were denied a hearing. The conference was held on a pile of cordwood outside the Queen of the Hills works, overlooking the town and the crowd. As neither side was prepared to compromise, nothing was accomplished.[49]

The bulk of the miners, however, had expected that some settlement would be reached. So when McDevitt reported back to them that no "arrangements" could be made, they reacted angrily. "The miners' committee had not completed their report," Picotte recalled,

when the cry of "$4 a day or blood" arose, and was instantly taken up and repeated by scores of voices. At the same time the flag of the Union was unfurled, guns were brought out, triggers cocked, and a column formed with the loudly avowed purpose of taking the mines and causing "blood to flow in the streets of Broadford and Bellevue." . . .

The column started, men "falling in" the ranks and swelling it at every step. Only 150 feet away was the "dead line." The head of the column advanced, the cooler Union men vainly endeavoring to delay it. The "dead line" was almost reached. The deputies at the Queen mine cocked their guns and prepared to take aim. But ten steps more and the carnage would have begun! It would have begun and ended no one knew where. At this moment Charles and Pat Furey and Acting District Attorney Hawley, who had vainly endeavored to prevent the forming of the column, succeeded in halting it, then in causing a break in the ranks, and finally in obtaining a parley.[50]

Sheriff Furey had succeeded in stopping the miners only when he agreed to have the scabs brought out of the mines.

While the miners crowded along the "dead line," Furey had Lusk bring his men out on the dump of the Queen where McDevitt and three other union men could talk to them. The union men told them simply, "We are trying to get $4 per day for you as well as for us. Look at that crowd of men down there trying to keep up wages. If you are in need we will assist you. Will you stay with us until this difficulty is settled, or not?" Some of the Queen men agreed to quit till the strike was ended; one, who complained that he had a wife and three children to support, was given $10 right then; another explained that he was a stranger and had just happened into the job; others said that they were already getting $4 a day from the contractors; and one argued that this was a "free country" and he would put his own price on his labor. But all quit at least for the day. The sheriff and the committee then proceeded to the Minnie Moore and all of the men there also quit.[51]

The sheriff's action was severely criticized; several of the more bloodthirsty deputies quit because he kept the miners from crossing the "dead line"; his bondsmen withdrew; and the county commissioners threatened to declare his office vacant. But he argued, "I did what I considered and still consider best, and will continue to do what is best, regardless of what croakers may say."[52]

The miners felt that they had won a victory and they headed home jubilantly. But that night the sheriff telegraphed acting-governor Curtis to send federal troops. The governor arrived at Bellevue the following morning, accompanied by General J. S. Brisbin, but without the troops. Like the enthusiastic miners who had arrived the day before, the governor and the general also vowed that they had "come over to settle this thing, and will not leave until it is." The Bellevue Safety Committee "demanded" that the governor declare martial law, but he refused. Picotte supported his stand, warning "Leadville has never been half the place since martial law prevailed there, and Tombstone is no better. Martial law or protracted strikes kill a camp for everybody."[53]

The governor instead set about organizing an arbitration committee to meet on March 22d. But Palmer and Lusk refused to

attend, snapping "we will listen to no dictation as to the conduct of our mines from the union or any other person." At this
the governor and the general left in disgust, concluding there
was too much "hot blood on both sides."[54]

The strike by then had brought business in Bellevue to a stand-
still and many merchants there were on the brink of bankruptcy.
The failure of the governor's arbitration attempt thus spurred
them to take direct action on their own to save themselves from
financial ruin. The next day the "stranglers" turned out in full
force, armed and deputized, to escort the scabs back to the
mines at Broadford. Several strikers tried to stop them, but they
were promptly arrested. At the same time Furey began arresting
McDevitt and other union leaders on charges of riot, stemming
from the St. Patrick's Day proceedings.[55]

Within a week nearly a hundred scabs were at work in the
Minnie Moore and Queen of the Hills at $3.50 a day. Picotte
saw the strike was clearly lost and called upon the miners to
concede defeat. But many of the union men were still determined to "hold out for a principle." Thus the strike technically
continued for two more months, although during this time the
Minnie Moore and Queen of the Hills ran uninterruptedly with
full crews of scabs. As there were no further demonstrations by
the miners' union, the guards were taken off the mines early
in April. Some of the scabs, still fearful, however, went to work
well armed. The only fatality of the strike, in fact, occurred
when a scared young Mormon teamster, scabbing on the Queen,
accidentally shot a hotel clerk while showing what he would
do if any of the union men tried to give him trouble.[56]

Finally on May 19th both the Broadford and Bullion miners'
unions conceded that they "have given up the strike on Wood
River, and will not attempt longer to stand out for $4 per day."
The strike had run 119 days, the longest yet in the western
mines, although the Tombstone strike was just two days shorter.
Once the strike was ended, the other companies on Wood River
also cut wages. An exodus of skilled miners again followed.
Some retreated to the $4 camps of Nevada, but most went on to
the copper mines of Butte.[57]

An important factor in the failure of the strike had been the shift of the apparent issue of dispute from wages to the contract system, undercutting much of the community support that the miners had on the wage issue. This course had been effective in breaking strikes in the western mines before, and with further refinement it became the principal strategy of the management in the remaining skirmishes of the $4 fight. That it was indeed only a strategy for breaking the strike and not a serious effort to eliminate the dispute over wages by adopting an alternative method of working the mines is obvious from the fact that in all cases the management promptly returned to the wage system once the strike was broken and a wage cut effected.

The failure of the strike and the resulting exodus temporarily broke the Wood River unions. The Broadford Miners' Union disbanded that summer and the Bullion Miners' Union folded in October. Within a year or two, however, the Broadford miners reorganized as a Knights of Labor assembly and successfully resisted a further wage cut in the Minnie Moore and Queen of the Hills in February of 1888.[58]

Soon after the commencement of the Wood River battle, the $4 fight also broke at Eureka, Nevada. Here the Ruby Hill Miners' Union, which was the strongest in Nevada outside of the Comstock, had vigorously resisted even the slightest infringement on the $4 wage. In the summer of 1884 that union had ordered miners to quit work in the Eureka Tunnel and the Ruby-Dunderburg mine because both companies were paying part of their wages in stock. From these actions it was obvious that any direct attempt to cut wages would inevitably result in a long and costly strike. Still several of the companies were determined to reduce labor costs, so following the example set at Wood River, they embarked on the more oblique strategy of first changing over their mines exclusively to the tribute system, by which the miners received a fraction of the value of the ore rather than a fixed wage.[59]

The shift from wages to tributing was made over several months in the Richmond, the Eureka Consolidated, the Ruby-Dunderberg, and the Bowman mines. Although some tributing had been done in the mines for years, the system was not popular and even a gradual shift to it did not pass unnoticed. One of the loudest opponents of tributing was the editor of the Ruby Hill *Mining News*, William J. Penrose, a Cornish miner and one of the founders of the Tuscarora Miners' Union. Claiming that some of the tributers averaged less than $2 a day, he called upon the miners' union to "put their foot down on this starving mode of working and make the fat officers of these large properties pay the men who go down into the mines regular wages." But the Ruby Hill union had already tried to abolish tributing in the Richmond mine in 1876 and had failed mainly from lack of community support. Community attitudes had changed little in the intervening years, so the union was reluctant to oppose the system again.[60]

But when superintendent McAulay of the Bowman went one step further and put on nonunion tributers in defiance of the closed shop the union finally took action. On March 23d, 1885, the miners went en masse to the Bowman, ordering the two non-union men to quit work immediately and "not to work again in any of the mines in Eureka County." In response McAulay stopped all work in the mine. The union men, considering the matter settled, held a "blowout" that evening to celebrate.[61]

The local press offered no criticism of the union action. But the mine managers quickly seized upon it as an opportunity to deal a crippling blow to the union, if, in fact, they had not provoked it for that very purpose. Rallying support from merchants fearful of the financial losses of a long strike, the managers organized the Eureka Citizens' Protective Association to defend the tributers and had the union officers arrested and jailed on charges of conspiracy. The union secured Pat Reddy of Bodie to aid local attorney Peter Breen in the defense. But the trial dragged on for months as all of the prospective jurors were disqualified when Reddy showed that the managers had tried to pack the jury. The case was finally transferred to Elko County before the miners were finally acquitted.[62]

In the meantime, however, the Protective Association had recruited seven paramilitary companies and issued a manifesto, recognizing the rights of the mine owners to make whatever contracts they wished and warning "all persons claiming the right to dictate to corporations or individuals in this community as to what course they shall pursue, to forthwith abandon such an unwarranted usurpation of power."[63]

With this support the mine managers discharged the remaining wage men and replaced them by tributers, many of whom were nonunion men. The union, with its officers still in jail, attempted no further interference. By the beginning of summer there were not twenty men working for wages in the entire district. As the directors of the Ruby-Dunderberg later boasted, "it was chiefly owing to the action taken by this and two or three other companies, in declaring not to employ miners any more on day's pay, and to do all their work either by tribute or contract, that the miners' union was broken up, the result being that miners' pay has been reduced from $4 to $3 a day." The broken union disbanded the following year.[64]

Within months of the defeat of the $4 wage at Eureka, wages came down in silver camps throughout Nevada. At Austin the Manhattan mine, the largest in the camp, cut wages to $3.50 on June 10th, although following the Wood River and Eureka plan the company had not employed any miners by the day for nearly two months. The Austin Miners' Union decided it was futile to resist and, rather than repudiate their pledge not to work for less than $4, they disbanded on June 18th. The Lewis union disbanded the same month, when the Betty O'Neal cut wages to $3.50 there. In September the Spring City Miners' Union also disbanded, donating their hall to the local school district, when the Paradise Valley mine dropped its pay by 50¢ a day. Most of the union camps even then, however, fared better than the nonunion camps. For at Candelaria in southern Nevada, where there was no union, wages were whacked clear down to $3 a day in July. The miners struck and attempted to organize a union but it was too late.[65]

"The day for $4 wages is past," one mine owner proclaimed, "and it is folly for anyone to speak of keeping wages up to that figure, because the mines cannot stand it. They must reduce wages and all other expenses, or shut down. The Miners' Union troubles in Eureka . . . and at Broadford were the last expiring gasps of a vicious and unreasonable kind of men."[66]

But that was not quite the end of it, for the defeat of the $4 wage had delayed, but violent, repercussions in the adjacent gold camps. An influx of miners from the Nevada silver camps flooded the labor market in the old gold camps of Nevada County, California, and some of the operators tried to use the situation to cut wages there to less than $3 a day. But they were met with determined opposition. The Nevada County Miners' Union, organized at Nevada City in November of 1887 with a branch at Grass Valley, mustered enough strength to block the wage cut. Whereupon the owners adopted the Wood River strategy of discharging their wage men and letting contracts only to nonunion men. Here the union, like those in Nevada, found itself powerless to resist. In the frustration of defeat a few men sought vengeance and, for the first time in a quarter century of western mining labor disputes, dynamite came into play. Hiding on the hill above the Providence mill, the amateur avengers slipped seven sticks of dynamite on a long fuse into the water pipe running to the mill. The bundle stuck in the pipe and fizzled out. But the would-be dynamiters tried again. This time they blew up the pipe and the resulting flood flushed tons of rock and mud into a neighboring mill. Both the owners and union men alike were outraged at the incident and raised over $1,000 reward for the capture of the "fiends." The era of the dynamiter had begun, however, and it was only a matter of time before he took a heavier toll.[67]

Thus the $4 fight came to a bitter end. Only on the Comstock and at Bodie did the $4 wage survive; on the Comstock the unions had compromised on longer working hours, while at Bodie the union's strength was so formidable that the owners apparently dared not attempt a cut. But wherever wage cuts were tried they eventually succeeded. For although community support for the miners on the issue of a wage cut was still strong, when the managers circumvented the issue by temporarily shifting from wages to contract or tribute work, the basic issue was no longer as clear and with the prospect of a long, ruinous strike facing them, much of the business community finally backed efforts of the management to bring a quick end to the dispute. The miners on the other hand had generally held to the simplistic demand for a nonnegotiable $4 minimum wage that had worked so well for so long on the Comstock. But it failed else-where in the West and in the wake of these failures lay the wrecks of several unions. Ironically it was the Comstock union that took the first steps toward negotiated settlements with a compromise on working hours. But the days of a more complex negotiated wage were still to come and would require conces-sions not just from the miners but from the mine owners as well.

The Gibraltar of Unionism

The decline of silver mining in the mid 1880s and the resulting blow to the mining labor movement was soon countered by the rapid rise of copper mining and the emergence of Butte, Montana, as the "Gibraltar of Unionism" in the western mines. Many miners who left the silver camps of Nevada, Idaho, and Colorado with the fall of the $4 wage and the breakup of the unions flocked to Butte, where, although wages were also only $3.50 a day, they found plenty of jobs in the rapidly opening mines and a new opportunity to revitalize the mining labor movement. By the end of 1885 Butte had become the most productive camp in the West, surpassing Leadville with a production of over $12 million a year and with increasing yield it maintained that preeminence well past the turn of the century. In 1885 also, the Butte Miners' Union became the largest in the West, surpassing the Virginia City union with nearly 1,800 members, and, growing with the camp, its ranks swelled to more than six thousand by the turn of the century. Butte's miners thus became "the foremost advocates of organized labor in the entire West," and Butte itself was boasted to be "the strongest union town on earth!"[1]

But the rise of Butte and its union was not an easy one; they grew by fits and starts with "many obstacles and difficulties to battle against." The camp was situated below Big Butte, from which it took its name, on the headwaters of Silver Bow Creek just west of the continental divide in one of the less scenic portions of western Montana. It had started as a placer gold camp in 1864, but attracted little attention until the opening of the first silver lodes in the Alice and Lexington mines in the late 1870s. The Butte Miners' Union owed its beginnings to the Walker Brothers' decision to cut wages to help defray the cost of pumps needed to open the Alice below the water level. The Walker Brothers, or more likely their superintendent Marcus

Daly, who later became a power in Montana, persuaded A. J. Davis of the Lexington to join in the cut.[2]

On Monday morning June 10, 1878, the two mines reduced miners' wages in the stopes and dry levels from $3.50 to $3 a day, although Daly still paid the old rate in the lower wet levels. The miners opposed the cut, and gathering support in the neighboring mines, they paraded the streets in protest, four hundred strong with a brass band. That evening at the Orphean Hall they decided to organize a union as "the most effective means of opposing the attempted reduction." Aaron C. Witter, a young miner from South Bend, Indiana, chaired the meeting. Three days later on June 13th, Witter's twenty-ninth birthday, the organization of the Butte Workingmen's Union was completed and Witter was elected president. The constitution and bylaws were "substantially those of the Miners' Unions of Nevada." The most significant difference, as the name implied, was the admission of all workingmen rather than just miners. This policy was staunchly defended by the editor of the Butte *Miner*, who argued, "If wages are reduced in the two mines in question, the reduction will very soon be general, all over the camp, and not only among miners, but with other classes of labor as well. In a mining country the daily wages of the miner is the standard by which the wages in all other departments of labor is regulated. To make the union successful, then, the entire employed, or wage-earning class must join it." Most workingmen heeded the call and within two weeks there were three hundred members on its rolls.[3]

The strike dragged on for over six weeks. But despite the diversity of its membership, the union maintained solidarity in their opposition to the wage cut and the peaceful course pursued by the strikers won them strong support in the community. The dispute was finally ended late in July when the union agreed to allow the reopening of the upper levels on contract, as long as the contractors paid at least $3.50 a day. Although mine managers elsewhere used the introduction of the contract system as an effective tactic in breaking strikes against wage cuts, the Butte managers were apparently sincere in seeking it only as a compromise measure to end the dispute, for they made no attempt to discriminate against union men in letting contracts. The editor of the *Miner* praised the union for the "quiet course and peaceful termination" of the strike. He considered this their greatest victory since it gave outside capitalists "convincing proof that Montana is a law-abiding country, one of the very safest in which money can be invested."[4]

But when the Lexington mill started up a few days later, he concluded that he had been a "little premature." Without notice Davis hired new mill men at the reduced rate. The union protested furiously. Threatening letters were sent to the $3 men, ordering them to "desist," and to Davis warning him that unless he paid $3.50 a day his life would "not be worthe hell." The union, repudiating the letters, sent a committee to try to work out a settlement with Davis.[5]

But at the union meeting on July 28th, when the committee reported back that no settlement could be reached, the union was undecided as to what to do next. After the meeting adjourned, however, a number of men concluded to take action on their own. At 11 o'clock that night five men visited the mill warning the workmen to knock off work within half an hour or "take the consequences." Davis hastily summoned the constable and a posse of citizens to stand guard at the mill till daylight. During the night several small groups of men were seen in the neighborhood of the works but no attempt was made to approach the mill. As soon as the posse had left in the morning, however, another "committee" visited the mill, ordering all $3 men to quit work immediately. This order was promptly obeyed, the men "not choosing to risk their lives for the chance of earning a living by hard work."[6]

This action turned some community support against the union, and an excited group met in the schoolhouse to form a Citizens' Protective Association. Some eighty-four "supporters of order" joined up, vowing to "protect from injury or interference any man who wishes to labor for any person or company on any terms." Despite its clearly stated purpose, the editor of the *Miner* tried to reassure the union that the association had "nothing to do with the settlement of the wages quarrel," as most of its members still opposed a cut. At the same time the union sought to reassure the citizens of Butte. "We are not Communists. Our object is to protect labor and laboring men's interests in a lawful and just manner." The formation of the Protective Association nonetheless brought a prompt end to protests against mill wages.[7]

The Workingmen's Union thereafter made no attempt to regulate the wages of any but underground miners and they ended the strike without even a clear resolution of that dispute. The continued growth and development of Butte as a silver camp over the next few years, however, soon stabilized underground wages at $3.50 a day without further action by the union. Through the growth of the camp the union also grew, boasting more than eight hundred members by 1881. In May of that year they reorganized as the Miners' Union of Butte City. This change formally recognized the fact that only the wages of underground miners were protected, but all workingmen were still eligible for membership and for sick benefits of $8 a week.[8]

But the prosperity of the union was short-lived. In the fall of 1881 the miners purchased a 50-by-100-foot lot on upper Main Street for about $1,600. "The deal was considered cheap," former president Pat Boland recalled, although an outcropping of bedrock made it "the most uninviting in the city, and any person but a miner whose courage is never questioned would pale at the idea of erecting a building upon the ground." The foundation and part of the lower floor had to be quarried out, but the rock taken up was used to raise the walls. The construction, however, became much more expensive than the miners had expected. The union treasury was soon exhausted, so they issued bonds that netted $10,000 and took a loan for an additional $4,000. The completed hall cost nearly $23,000. Severe weather also slowed construction during the winter, but by February of 1882, the hall was at last nearly finished. The miners were happily preparing for a grand opening ball, when the whole building collapsed.[9]

The collapse of the union hall nearly brought about the collapse of the union. The disillusionment and the debts incurred caused many union members "to become lukewarm and in bad standing." The union reached its nadir the following spring.

When Pat Boland was elected president in March of 1883, he later recalled, "we had enrolled and in good standing the Spartan band of seventy-eight and the magnificent sum of 45 cents in the treasury, not a very tempting amount to flee to Canada with." But despite their destitute condition the union voted to increase sick benefits to $10 a week and to make the financial secretary a paying position at $3.50 a day. "This," and perhaps his own election, Boland felt, "was the turning point for the better in favor of our noble organization."[10]

But the rapid development of Butte's newly opened copper ores, whose yield surpassed its silver output that same year, was doubtless much more effective in revitalizing the failing union. The new jobs opened up in the copper mines came at an opportune time for many miners elsewhere in the West where declining silver prices were closing mines and breaking up unions in the $4 fight. These new jobs attracted an army of militant union men from the old silver camps and this "new blood . . . gave the union renewed life."[11]

With this new blood, however, came bad blood between the Irish and Cornish miners. Prior to the break of the $4 wage in the silver camps, the lower wages at Butte had attracted mainly Irish and American miners who had been unable to compete with the Cornish for better wages elsewhere. But with the reduction of wages in the silver camps, many Cornish miners also came to Butte, and a competition for jobs and control of the miners' union began. Coupled with the traditional antagonism between the English and the Irish, it would slowly grow to a dispute of grotesque proportions. But for a time at least this cancer festered only in the background and the union made giant strides.

In just two years the Butte union's membership swelled to nearly 1,800, surpassing even the Virginia City union and making it the largest miners' union in the West. With this resurgence the union paid off its indebtedness and once again began construction of a hall. This time the miners' efforts were not frustrated. The rubble of the old hall was cleared away and work commenced in the spring of 1885. The hall was completed that fall. It was an impressive, substantial structure measuring 50 by 98 feet with two stories—the lower of stone, the upper of brick. The meeting hall was on the upper floor with a ladies' sitting room, reception room, parlor, and offices, all "well and tastefully furnished" at a cost of $3,000. The building itself cost $13,000 and was paid off within a few years from union dues and rents from the hall itself. The miners' union met in the hall every Saturday night and rented it out to other labor and fraternal organizations the other nights of the week. The ground floor was divided into two store fronts that were also rented out as everything from a beer hall to a haberdashery and a post office, contributing another $300 a month to the union's treasury[12]

With the development of the copper mines in Butte, the labor force had become so large and so diverse that the miners finally felt they should restrict their union membership solely to mine workers. Thus in March of 1885 the union reorganized once again, limiting membership to miners and streamlining its name to the Butte Miners' Union. Other workers were encouraged to join local assemblies of the Knights of Labor or to organize their trades.[13]

To aid such organization and give solidarity to the labor movement at Butte, the Miners' Union joined with the Knights of Labor and the Typographical and Tailors' unions to form the Silver Bow Trades and Labor Assembly on January 2, 1886. With the prodding and protection of the Assembly many other trades were soon organized. By 1891 there were thirty-four affiliated unions, representing nearly all of the six thousand craftsmen and laborers in the district. Stephen C. Graney of the miners' union was the first president of the Assembly and the miners, as the largest union, exerted a strong influence over its policies. A new Workingmen's Union organized in May of 1890 by the surface workers and other laborers became the second largest union with 1,800 members. These two unions gave enormous strength to the Assembly and all of the smaller unions through their solidarity of action, making Butte truly a bastion of unionism.[14]

Following their reorganization along somewhat stricter lines, the miners' union began to press for a closed shop in the mines. One by one and without incident the nonunion miners were in-

duced to join up until, by June of 1887, the closed shop had been
established in all the mines but the Bluebird. A confrontation
came on June 13th, Miners' Union Day, when superintendent
Booraem refused to shut down the Bluebird for the day, as all
the other mines had in honor of the union's ninth anniversary.
That morning the union men marched out to the mine to "gently
intimate to the men in charge that the shutting down of the mine
would be in accordance with the eternal fitness of things." Some
six to eight hundred men fell into line behind the flag, the Alice
and Emmet Guard bands, and a carriage, bearing the speakers
of the day: union president Samuel Barker, Butte *Mining Journal*
editor and former miner William J. Penrose, Methodist minister
W. E. King, and labor organizer John M. O'Neill.[15]

Just as the noon whistle was blowing the procession reached
the Bluebird. Penrose, as spokesman, demanded that Booraem
shut down the mine, but he refused. When he further refused to
let the union men enter the engine room, Penrose recounted,
"someone in the rear noticed a piece of rope, about three or four
feet long, lying near the trestle. In a spirit of mischief it was
taken up, tied in a noose and carelessly thrown into the air, and
alighted, surely by accident, upon the head and shoulders of the
superintendent." At this Booraem suddenly decided not to
offer further resistance. The union men brought up the crew,
all Italians, and marched them back into town. After parading
their captives up and down the main streets they were taken to
the union hall and initiated into the union. The miners were
fired with enthusiasm over the day's victory and with speeches,
band concerts, and a grand ball they celebrated till nearly daylight the next morning.[16]

The Butte press was unanimous in its praise of the union on
its anniversary, but made no editorial comment on the Bluebird
incident. The tenor of the praise was set by the *Inter Mountain*
editor, who lauded the union as "the most independent, most
orderly, temperate and prosperous body of workingmen in the
world."[17]

The outside press, however, was not so favorably impressed.
Rossiter Raymond of the New York *Engineering and Mining
Journal*, visiting Butte the following month, condemned the

Bluebird incident as "an un-American outrage as well as a blunder." "And," he added,

the Miners' Union of Butte has been carrying matters with a high hand, in a way to suggest that the spirit of the Mollie Maguires, hunted out of Pennsylvania by the sword of justice, has taken refuge and gathered fresh courage in the Rocky Mountains. I do not mean to say that any of the fugitive criminals who are "wanted" in Schuylkill County are masquerading now in the disguise of champions of labor at Butte, but I do say that the Butte Miners' Union is coming dangerously near to their methods of violence and terrorism.

Raymond was further struck "with dumb astonishment" to learn that William Penrose, "one of the most prominent abettors of this outrage . . . who had insulted public decency, imperiled public safety and defied public justice with his incendiary ribaldry," had been appointed by the Governor as the impartial and disinterested third member of a newly created Territorial Board of Arbitration. "Strictly speaking," he quipped, "he is eligible under a certain construction of the law; for nobody pretends that he represents capital, and Heaven forbid that he should be considered as representing labor!"[18]

Penrose accused Raymond of reviving the "long-settled episode" at the Bluebird only as a pretext for an "insidious attack" upon Butte, and he suggested, "Mr. Raymond can go to hell." Raymond snapped, "This appears to be a kind of free pass, issued by an agent of the line," and concluded that the episode was only " 'settled,' like any other nasty precipitate, to rise again on the first agitation of the waters. The fact is, mere 'settling' wont do for PENROSE. He ought to be filtered out and thrown away."[19]

For all his wit, however, Raymond had little effect in Butte, and Penrose on the political patronage of the Miners' Union went on to win a seat in the legislature. Although not then a member of the union, as a former union miner, he gave the union strong support in his paper. Born in Cornwall in 1856, he had started work in the tin mines at the age of seven; coming to America he dug coal in Pennsylvania and Illinois and silver ore at Tuscarora and Ruby Hill before he quit the mines to become a newspaper reporter and eventually the editor-proprietor of the Ruby Hill *Mining News*. He was an active union man throughout and with the collapse of the Ruby Hill Miners' Union in 1885 he joined the exodus to Butte where he started the *Mining Journal*. There he took a conspicuous part in furthering the union's cause, seeking in return its support as a political base. To help bring in the miners' vote on the Democratic ticket, he was given that party's nomination to the state house of representatives in the fall of 1889. He won handily and ran for a second term the following year.[20]

Here, however, Penrose came into conflict with Patrick F. Boland, an energetic and popular Irishman, and the three-term past president and prime mover of the miners' union, who got the Republican nomination for the state senate with the expectation that he could deliver the miners' votes on that party's ticket. Underlying this conflict was the growing antagonism between the Cornish and Irish miners in Butte. As the campaign grew heated Penrose unleashed a personal attack on Boland and his administration of the union. He denounced Boland and his friends as "professional labor men, who never work except to 'work' the genuine laborer." In the union election that September he openly backed challengers to "Boland's gang" and made much of the defeat of two of his Irish "henchmen," Eugene E. Kelly and William E. Deeney who lost their seats on the Trades and Labor Assembly. At the polls in November Penrose won reelection as the Democratic ticket generally carried the day, but Boland lost his election by only a slim margin of about a hundred votes out of some six thousand. The bitter feeling generated by the campaign erupted again in a tragic dispute over the eight-hour issue.[21]

The eight-hour fight in Montana had begun at the state constitutional convention in 1889, when the Committee on Labor, led by Aaron Witter and Peter Breen of the Butte Miners' Union, tried unsuccessfully to put an eight-hour limit on all state work. The following year, when copper and silver prices were up again, the Butte miners tried more direct action. On November 21st the union called upon the mining superintendents to establish eight-hour shifts, arguing, "we believe that the miners of this district if subjected to the inhalation of poisonous powder fumes for a shorter period than at present will be able to perform proportionately more work and that the interests of the companies will not suffer by a reduction of the hours of labor." Penrose quite uncharacteristically made no comment on the issue. When the superintendents rejected the request, the miners again turned to the legislature.[22]

Boland and other miners' union officials drafted a bill to make the eight-hour day mandatory in the mines. It was introduced in the house of representatives in January of 1891 by Peter Breen. Entitled "A bill for an act to regulate contracts for underground labor and to provide for the enforcement thereof," it specified:

Section 1. It shall be unlawful for any person to become a party to any contract for the underground service or labor in mines of one and the same person for more than eight hours in twenty-four consecutive hours.

Section 2. Any person violating the foregoing action shall be fined not less than $10 nor more than $100, or imprisonment in the county jail for not more than six months, or both such fine and imprisonment.[23]

The bill stirred up a hot debate. Its opponents insisted that the legislature had no right to regulate the working hours of any but state employees. Its supporters, however, argued that it could be passed as a "sanitary measure." The Butte Miners' Union president, William Eddy, went to the legislature to personally lobby for the bill and hundreds of miners in outlying camps sent in petitions supporting it. The Butte mining superintendents openly petitioned for its defeat.[24]

Throughout the early debate on the floor of the house, Penrose remained silent, but in the columns of the *Mining Journal* he suggested mass meetings in all the camps for a "full and fair expression of opinion ... followed by united and intelligent action." Although he had remained noncommittal, the bill's supporters had still assumed that he would vote for it. Thus when it finally came up for a vote in February they were furious when he suddenly took a strong stand against it, warning

passage of the bill would surely cause many mines to shut down and consequently cause the discharge of thousands of miners; . . . eight out of every ten business houses would be closed up and Montana as a mining center would sink to the level of the copper centers of Vermont and Michigan, the iron customs of New Jersey, the coal methods of heartless Pennsylvania, the craving practices of the gold kings of California and the domineering rules of the moneyed kings of the Old World.[25]

The bill was defeated by a decisive vote of 31 to 21, which split both the Butte votes and party lines. The defeat was greeted with "anger and disgust" by the miners in Butte and the union meeting turned into an "indignation meeting" to condemn "those unscrupulous and unreliable individuals who were elected by the working masses of Silver Bow . . . only to grossly betray the trust thus reposed in them when the test came." They were branded the "willing tools and serfs of corporations . . . the lackeys of the millionaires and the hacks of the big corporations." As a constant reminder of their treachery, a giant blackboard was hung in the union hall, inscribed with their names and above them was written "TRAITORS AND ENEMIES OF LABOR." Heading the list was "W. Judas Penrose." The Butte Workingmen's Union, the Typographical Union, the Butchers' Association, and the Carpenters' Union all joined with the miners in denouncing the opponents of the bill and Penrose in particular.[26]

Penrose responded with bitter new attacks on Boland, Deeney, and Breen, and with caustic burlesques of union affairs through the imagined proceedings of the "Chippies' Protective Union" and the "Butte Perpetual Rest Society." To answer these attacks the Silver Bow Trades and Labor Assembly finally started a newspaper of its own, the *Sunday Bystander*, on March 22d.[27]

The fight was still raging when shortly before midnight on June 9th Penrose was shot dead on a street corner in Butte. Most of Butte denounced the "dastard deed," although there were many happy to see him gone, and the press called it "the cruelest and most-cold-blooded assassination ever known in Montana." The Butte City Council, Governor Toole, and Senator Clark each offered a $1,000 reward for the arrest and conviction of the murderer. One editor, feeling that the reputation of the state itself was at stake, demanded that "every man should constitute himself a committee of one to hunt the assassin to his death at the rope's end."[28]

The police immediately arrested one Belle Browning, alias Emma Turner, who was said to have been insanely jealous of Penrose and caused him much trouble, sending "obscene letters" to his wife, stabbing another woman on his account, throwing rocks through his window and threatening to kill him. But a variety of other theories also circulated that Penrose had been murdered by some thug trying to rob him, by some enraged husband, by someone he had attacked in his paper, by someone he was going to attack, or by some secret organization "like the notorious Mollie Maguires."[29]

After two weeks of hearings the coroner's jury concluded his murderer was unknown and Miss Browning was released. A half dozen Pinkerton detectives were then put on the case and a month later they arrested three prominent union men for the murder. The three, Eugene E. Kelly, former president of the miners' union, Philip J. Hickey, former president of the Trades and Labor Assembly and treasurer of the miners' union, and William E. Deeney, president of the Workingmen's Union and former financial secretary of the miners' union, had all been among those whom Penrose attacked in the *Mining Journal.* But they all had alibis for the night of the murder and were acquitted after a lengthy hearing and trial, which lasted for months and cost the county well over $25,000. No one was ever convicted for Penrose's murder. Ten years later the miners and smeltermen finally won an eight-hour day in the state legislature.[30]

With its emergence as the most powerful union in the western mines, the Butte Miners' Union displaced the Comstock unions as the center of the western mining labor movement, and it began to take an active part in organizing miners elsewhere in Montana and aiding those beyond. But unlike the Comstock unions, the Butte union kept more paternal control over the new unions, organizing them first as "branches" before granting them independence. The first of these branch unions was started on September 28, 1888, at Granite, the second largest camp in Montana. With the aid of the Butte union, the branch won a closed shop in January of 1890 and signed up nearly a thou-

sand men. On September 30th of that year the Butte union
granted it an independent "charter" with jurisdiction over Deer
Lodge County. But the influence of the Butte union remained
strong. The Granite miners continued to celebrate the Butte
Miners' Union Day, June 13th as well as their own anniversary
and when they built a union hall in 1891 it was modeled in almost every detail after the Butte hall. Other branches were organized in the silver camps of Barker, Castle, Champion, and
Neihart. Each was ultimately given independence with a district or county charter. Their ties with the Butte union and each
other were maintained, however, and on January 15, 1892, they
all affiliated to form the Montana State Association of Miners, as
the Nevada unions had done some years before.[31]

But this association was only a first step toward the formation
of the much stronger and more broadly based Western Federation of Miners the following year. For at the same time the
mining labor movement was experiencing a revival elsewhere
in the Rockies. The cuts in wages and other operating costs in
the early 1880s had more than compensated owners for the
further decline of silver prices and brought for a time, at least,
a measure of prosperity in some of the richer silver-lead mines.
With this prosperity came the formation of a number of new
unions. In the Coeur d'Alene mines of northern Idaho the first
union was organized at Wardner on November 17, 1887, followed by others in the neighboring camps of Burke, Gem, and
Mullan in October and November of 1890. These unions with
a combined membership of more than two thousand were the
strongest and most militant outside of Butte and they were soon
to become the center of one of the bitterest mining wars in the
West. Other unions were organized in Colorado at Leadville in
1885, at Aspen and Breckenridge in 1886, at Red Cliff in 1887,
at Central City in 1888, again at Leadville and Aspen in 1890,
at Red Mountain in 1891, and at Creede, Ouray, Rico, and
Telluride in 1892, and also in Utah at Eureka and Mammoth
in the Tintic district about 1890. Many of the Colorado unions,
however, were assemblies of the Knights of Labor and followed
their more conservative policies against strikes. But in the few
strikes that did occur during this period, the miners generally

won easy victories. These disputes centered on hospital dues, alien labor contracts, company boardinghouses, and "pluck-me stores" as well as wages. Although the unions outside Montana had no direct affiliation with the Butte union, the Butte miners always stood ready to aid them in their struggles and when the collapse of silver prices in the early 1890s brought sweeping wage cuts throughout the West the Butte union rallied many of the unions to form the Western Federation of Miners for more effective resistance.[32]

The very success of the Butte Miners' Union, however, was at the same time the root of its greatest weakness. For with its phenomenal growth in number and power had come the growing struggle for the control of its offices and its political patronage, between the Cornish and Irish. The feud between Penrose and Boland gave only a shadow of the disputes to come, disputes that ultimately threatened to destroy everything that the union had built, and in fact would lead to the dynamiting of the union hall, the splitting of the union itself, and the near destruction of the Western Federation of Miners. The "Gibraltar of Unionism" was thus a shaky rock at best, but for a time at least it was strong enough to successfully unite all of the western miners' unions—a goal that the earlier unions had never even attempted and an achievement that would stand as a lasting monument to its service to the labor movement.

The Start of the War

The disastrous decline in silver prices in the early 1890s spurred new wage cuts in the western mines, triggering widespread lockouts and strikes in the bitterest and most violent mining war the West was yet to see—a war that would last for over a decade and eventually engulf nearly every major camp in the West. But the decade opened with bright promise for the western mines. The Sherman Silver Purchase Act of July 1890, raising the government's purchase commitment to 4 million ounces of silver a month was greeted with jubilation. Silver prices leaped from a low of 92¢ an ounce to a high of $1.21 that August. Many mines that had been forced to close with the earlier decline of silver in the mid-eighties were reopened and a new boom was proclaimed. The increased production stimulated by higher prices, however, far exceeded the government's purchase limit. By the summer of 1891 the growing silver surplus started prices on a steep decline. Silver was back down to 94¢ an ounce by the end of that year and fell to a new low of 84¢ the following summer. It held there for nearly a year before the general financial panic of 1893 and the repeal of the Sherman Act sent it plummeting again to barely 60¢ an ounce by March of 1894.[1]

Lead and copper prices were also dragged down by silver and the entire western mining industry was seriously crippled. One by one, as prices declined, the mines ceased to pay a profit. The poorer mines were forced to close down permanently and even many of the richer mines suspended at least temporarily while the owners sought new economies. A few mine owners attempted to find relief in lower railroad rates or in the introduction of machine drills and other more efficient mining techniques. But for the majority of mine owners and directors the answer was simply to cut miners' wages again and break up the unions if necessary. The miners, still suffering from the loss of the $4 wage, reacted bitterly and a full-scale industrial war erupted. The first violence flared in the Coeur d'Alene district of northern Idaho. There the owners pushed the attack with every weapon in their arsenal: midwinter lockouts, court injunctions, hired gunmen, Pinkerton spies, mass arrests, and state and federal troop intervention. When the miners responded in kind, the result was a violent conflict that paralyzed the district and plunged it into a state of guerrilla warfare for a decade.

But the first wage cut was made not in the Coeur d'Alene but at Candelaria in the sagebrush country of southern Nevada. The basic issues were the same as those that would later bring violence, but here, because the miners were poorly organized, a less aggressive policy on the part of the owners carried the day. In one of the last skirmishes of the $4 fight in September of 1885, the Holmes and Mount Diablo mining companies had broken a hastily organized union and cut wages from $4 to $3 a day. But, when silver recovered in August of 1890, they grudgingly agreed to a petition by their workers to raise wages to $3.50. The directors soon regretted this concession as silver once more began to fall, and in November of 1891, when it was back down to 95¢, they decided to cut wages back to $3 again. They threatened to close down if the miners rejected the cut. But to soften their demand a bit, they offered a faint hope of future restoration in promising to adopt a sliding wage scale tied to bullion prices. With this scheme if silver recovered to more than $1.07 an ounce wages would again be raised to $3.50 a day and should it ever go above $1.29 they would even pay $4 a day. They did not mention at what price they would cut wages to less than $3.[2]

The miners, meeting on November 27, rejected the scheme. Resolving that no miner be allowed to work for less than $3.50 a day, they organized the Candelaria Miners' Union. Both companies closed down four days later, and the directors in San Francisco signed an agreement that "under no circumstances" would they recognize the union. More than four hundred men were thrown out of work. The lockout dragged on for four

months through the long dreary winter and many miners left.
Only the moral and financial support of the Comstock unions
sustained those who remained.[3]

Finally even the most determined conceded defeat. On March
29th, 1892, they announced that they would accept the cut "in
view of the continued depression in the price of silver." But
even though the miners had conceded all that the company
originally demanded, the directors now sought revenge for the
miners' impertinence is not accepting the demands immediately.
They now refused to reopen the mines until the union disbanded.
The miners at last conceded even this point; on April 5th they
dissolved the union. But the promised resumption of work still
did not come for another three months, after most of those active
in the union had finally left the camp.[4]

The dispute in the Coeur d'Alene grew out of similar circumstances and similar breach of faith to that at Candelaria.
But the much more aggressive and provocative course pursued
by the owners here pushed it to a violent eruption. The silver-
lead lodes cropping from the slopes of the Bitter Roots along
the south fork of the Coeur d'Alene river had been opened in
the mid-1880s. By 1890 more than four thousand people were
crowded into the half dozen towns strung along the railroad
running through the narrow canyons—from the mines at
Wardner on the west, through the supply towns of Wardner
Junction and Wallace to the mines at Mullan on the east, and
at Gem and Burke up Canyon Creek on the north. The principal
mines were the Bunker Hill & Sullivan, at Wardner, the Gem
and the Frisco at Gem, and the Tiger & Poorman at Burke. By
1890 they were producing over $4 million a year, paying hand-
some dividends to their owners, and making the Coeur d'Alene
the principal lead-producing region in the country. At the same
time the miners in the district finally began forming unions to
demand a minimum wage.[5]

In the fall of 1890 the miners revived an earlier union at
Wardner and formed new unions on up the canyon at Gem,
Burke, and Mullan, all adopting the constitution and bylaws of
the Gold Hill union almost verbatim. To strengthen their po-

sition, they consolidated on January 1, 1891, under a Central
Executive Committee as the Coeur d'Alene Miners' Union. Immediately they began a double-pronged offensive to establish
both a hospital for the care of the sick and injured and a $3.50
minimum wage for all men underground. With broad community support the Miners' Union Hospital opened its doors
in May and is still operating today. The minimum wage, however, was not won so quickly. Miners were already getting $3.50
but muckers and carmen, many of whom were miners only recently displaced by the introduction of machine drills, were getting only $3. Taking on the mines only one at a time the union
made its demand. Many of the owners readily agreed; the others
did so after brief strikes. Thus slowly but deliberately the union
won the point. The last and largest mine, the Bunker Hill &
Sullivan, acceded to the minimum wage in August of that year.[6]

But a number of the owners who had agreed to the demand
only grudgingly quietly vowed to destroy the union. Led by
Alfred M. Esler, manager of the Helena & Frisco company, and
John Hays Hammond, president of the Bunker Hill & Sullivan,
they formed the Mine Owners' Protective Association of the
Coeur d'Alenes, which soon came to be known simply as the
MOA. Hammond was president and John A. Finch of the Gem
mine was secretary. Although the association also served to
represent the owners in rate negotiations with the railroads, it
was evident from the beginning that its principal aim was to
break up the Coeur d'Alene Miners' Union.[7]

The MOA quickly took the offensive against the union in the
summer of 1891, hiring Pinkerton and Thiel detectives to infiltrate the union as spies and perhaps as agents provocateurs.
A couple were soon discovered by the miners and drummed out.
But at least one succeeded very well. He was Pinkerton detective
Charles A. Siringo, who was hired to infiltrate the Gem union
in September. Under the alias of C. Leon Allison, Siringo was
given a job in Finch's Gem mine. He joined the local union,
quickly won the confidence of his fellow workers, and in December was elected recording secretary. "This," he boasted, "was
a useful position from my point of view, since it would give me

access to all the books and records." Although sworn to keep
all union matters secret Siringo reported every detail of the
union's affairs to his employers and generously branded the other
union officers as "dangerous anarchists, who had completely
duped the hard-working miners and were formulating demands
to which the owners could not possibly agree." Heading the list
of these sinister plans was abolishing the company store! His
conspiratorial employers were perhaps spurred on all the more
by the notion that there was a conspiracy against them. Hammond, in fact, seemed willing to believe almost any conspiracy,
no matter how fatuous, for he was left the rather bizarre notion
that the whole dispute was caused by "exasperated mine operators" in Butte who "diverted the attention of the agitators to
the Coeur d'Alene." Nonetheless Siringo served the MOA well,
enabling them to outmaneuver the union on several occasions.
Even his eventual exposure would work to their benefit, pro-
voking violent repercussions that would further weaken the
union and lead to state and federal intervention.[8]

But that fall the MOA got a more immediate opportunity to
greatly weaken the union. It came unexpectedly in the decision
of the Northern Pacific and Union Pacific railroads to increase
the rate for shipping ore to the Omaha and Denver smelters from
$14 to $16 a ton. Soon after the increase went into effect the
mine owners issued an ultimatum to the railroads, threatening
to suspend shipment if the old rate was not restored. The railroads refused and on January 15, 1892, all the major mines and
mills in the Coeur d'Alene effectively closed down. The mines
stopped extracting ore and only modest development work was
continued, employing but a handful of family men whom the
owners considered "loyal." Altogether about two thousand miners and millmen were thrown out of work. With the worst of
winter yet to come and no settlement with the railroads imminent many of the single miners began drifting off to seek work
elsewhere.[9]

Shortly after the closing the suspicion was openly expressed
that the MOA was using the shutdown to weaken the union and
that they would resume only at reduced wages. This was fully

confirmed on March 18th, when the MOA announced that the railroads had at last agreed to restore the old shipping rate. They declared that all mines would resume work about April 1st, but added, "Believing most earnestly that the advance of the wages of carmen and shovelers which was forced upon the mine owners during the past year was unreasonable and unjust . . . the association begs leave to announce the following scale of wages: For all miners, $3.50 per day of 10 hours; for carmen and shovelers, $3 per day of 10 hours."[10]

The miners, denouncing the owners' duplicity, rejected the wage cut. The battle lines were now clearly drawn. The shutdown had reduced the union strength to barely eight hundred, but those who remained were a solid corps of veteran union miners, many of whom had weathered more than one strike. Thomas O'Brien, president of the union's central committee, and Joseph F. Poynton, its secretary, were both veterans of the Comstock unions and Poynton had been one of the founders of the Bodie union.[11]

The MOA kept the offensive, publishing a lengthy, almost hysterical, tirade against the union. First they complained, "There was peace and amity everywhere until the fall of 1890 when the miners' union was first formed and trouble began almost immediately and we have had lots of it since and it has been increasing in volume and violence up to the present time." Then they blasted the union leaders as "agitators, who . . . desire to terrorize the whole community . . . and keep up a continuous state of turmoil and strife," and accused them of "threats of running the mine owners out of the country, of burning their mills, of blowing up their flumes, of even murdering them, shooting them in their beds and so on." Finally they asked,

Suppose under this state of affairs the mine owners were to accede at once to the ultimatum of the central committee, can it reasonably be expected that any permanent peace in our fair county would be secured? Vain hope! We venture the belief, and we have good grounds for it, that sixty days would not elapse until trouble would arise somewhere and the fiat of the union would go forth and must be complied with or the offending mine, or all of them, would be closed down and the men driven out. Under this state of affairs we, the mine owners, think it is about time to call a halt and we do so most emphatically; and we think every right thinking man in the country, be he miner or businessman, will say, "Amen. You are right and we are with you."[12]

This blast backfired on the MOA, for its outrageous exaggerations alienated many in the community who might otherwise have supported their position. Mass meetings of "right thinking" miners and businessmen, held at Wallace, Burke, Gem, Mullan, and Wardner, passed resolutions branding the MOA's charges as "palpably false and uncalled for and an unmerited insult to the whole community." Even the Spokane *Review* which had consistently sided with the owners criticized the tirade as "too dogmatic and combative."[13]

But in response to such criticism the MOA only hardened their position, announcing they would let their mines lie idle until June. And when they did resume, Esler let it be known in no uncertain terms, "We will never hire another union man." Thus the shutdown against the railroads now formally became a lockout against the union. The union made the closure of the mines complete by asking the men still working on development work to quit, which they did.[14]

In preparation for the resumption of work, agents for the MOA began looking for scabs in the Midwest and, to prevent union interference, they also sought injunctions against the union. Since the sheriff and most other local authorities were either union members or sympathized with the miners, the MOA provoked an incident in a mine incorporated outside the state, so that they could appeal to a federal judge and have the support of federal marshals, bypassing the local officers entirely.

On April 29th four scabs were sent to work in A. B. Campbell's mine at Burke—ironically named the Union. The miners responded predictably. Joined by many townspeople, they escorted the scabs down from the mine to a mass meeting at the union hall. After heated discussion two of the scabs recanted and joined the union. But when the other two refused, the crowd shouted, "Send them up the canyon! Burke is no place for scabs!" With a tin can serenade they sent them on their way toward the snowy crest of the Bitter Roots and the Montana state line.[15]

A week later with affidavits from the two exiled scabs, the MOA's attorney, Weldon B. Heyburn, obtained sweeping injunctions from the U.S. District Court in Boise. The injunctions named not only the Burke union, but also the Wardner, Gem, and Mullan unions, the Central Executive Committee, 120 named individuals, including the Shoshone county sheriff, lawyers and editors sympathetic with the union, and a large number of John Does and Richard Roes. They were all enjoined from interfering in any way with the company's operations or employees. James H. Hawley and two other attorneys for the union tried to have the injunctions set aside, but the court battle proved lengthy and unsuccessful.[16]

Armed with the injunctions, the MOA launched an immediate and massive assault. They had already advertised in the iron and copper mines of northern Michigan for 250 miners at $3.50 a day and they had sent one of their detectives, Joel Warren, into the farm towns of the Palouse to recruit guards at $3 a day. So on May 12th, the day the federal marshals began serving the injunctions in the Coeur d'Alene, the MOA shipped two coach loads of scabs by rail from Michigan and dispatched a special car of 55 armed guards east to rendezvous with them. There were a total of seventy-three scabs, mostly newly arrived immigrants from Austria, Poland, and Sweden. Many spoke no English and were unaware that they were going as strikebreakers.[17]

When word of these movements reached the Coeur d'Alene, Sheriff Cunningham obtained a warrant for the arrest of Joel Warren and the guards for violation of the state law against bringing an armed force into the state. He and the union men prepared to meet the train at Wallace. The Butte miners also planned to meet the train when it came through Butte to try to talk the scabs into turning back. Siringo, however, tipped off the owners to these preparations. The two coaches of scabs were separated from the regular train before it reached Butte and a special train was made up with just these two cars and the car carrying the guards. With the scabs locked into their cars, the special bypassed both Butte and Wallace. At Butte the miners boarded the regular train only to find the scabs were gone and at Wallace the sheriff with a huge crowd waiting at the depot watched in frustration as the train roared through a switch just short of them and up the side canyon toward Burke without stopping. The train finally halted below Burke, where the scabs and guards scrambled up to the Union mine, a thousand feet above.[18]

At the mine the scabs were put under the command of none other than George Daly's old lieutenant, Joe McDonald, who had fought the miners' unions at Bodie and Leadville over a decade before. McDonald had barricaded the mine and stock-piled supplies for a long siege. One of the guards was deputized as a federal marshal and stationed at the foot of the trail leading up to the mine. He stood ready to serve an injunction on anyone who approached the mine and, when the sheriff arrived, he signaled the mine to have the guards hide their guns and pretend to work. As a result, by the time the sheriff got up to the mine the only man he could identify as a guard and arrest was Joel Warren. The MOA promptly bailed him out on a $2,000 bond.[19]

As soon as they realized that they had been brought in as scabs, many of the men demanded tickets back home. When McDonald refused, they began to desert. Twenty-seven slipped away within the first three days, leaving more guards than miners. But their places were soon filled, for this was only the first wave of the assault. More carloads of scabs and guards followed in quick succession to reopen the Frisco, the Gem, the Bunker Hill & Sullivan, and several other mines. Within two weeks the owners claimed to have three hundred men at work and by the end of June eight hundred.[20]

Nonetheless even this massive assault failed to break the union. Although they were restrained by the injunctions and intimidated by the guards, the union miners still succeeded in convincing roughly half of the scabs to quit and join with them. As part of the inducement the "converts" were given the same financial aid as other union members. This aid came primarily from neighboring miners' unions and other labor organizations. The Butte union assessed each of its members $5 a month to aid the Coeur d'Alene miners and the other Montana unions and those in the Black Hills did almost as much, providing nearly $30,000 a month in aid. The Spokane labor committee raised additional funds and sent freight car loads of food and supplies gathered with the aid of the Farmers' Alliance. Union president Tom O'Brien also obtained a $10,000 firewood contract which gave many of the miners temporary work at $3 a day. But even with all of this support the hardships were severe. By the end of May the dispute had become the longest in the history of the western mines and it showed no signs of ending. A sense of frustration verging on desperation was slowly growing among the miners.[21]

Although they had succeeded in opening the mines on their own terms, the MOA was still not happy with the situation either. The injunction had not broken the union and the cost of maintaining guards at the mines made operations unprofitable. Thus they apparently concluded the only solution was the imposition of martial law, which would at least relieve them of the cost of guards and might also be effective in crushing the union.

Late in May the MOA circulated a sensational story that union men had held up a train at Mullan, and, charging that the union had "jeopardized" property, "terrorized" people and "set the laws at naught," they called upon Governor Norman Willey to declare martial law. Instead Willey visited the Coeur d'Alene to investigate the charges firsthand. Although he was wined, dined, and entertained by the mine owners, he concluded that the alleged holdup was a gross exaggeration, so he refused to call in troops.[22]

But to appease the owners somewhat, he did issue a proclamation on June 4th ordering "all unlawful assemblages, combinations and bodies of men to immediately disperse." This prompted a new round of mass meetings in the Coeur d'Alene to denounce the implied charges against the union. The governor, sensing the volatility of the situation, also called upon President Harrison a few weeks later to station federal troops in the district as a deterrent to violence. The president's refusal was widely publicized on July 5th.[23]

It was now obvious that nothing short of a truly violent outbreak would prompt the government to bring in troops—the vague exaggerated charges and political influence which had worked so well at Amador and Leadville were no longer enough. Just such an incident did, in fact, occur within a week. To what extent it was overtly, or covertly, triggered by the MOA's guards, or by its undercover detectives acting as agents provocateurs within the union, can only be debated. Clearly the MOA had much to gain and the miners' union had much to lose by the incident. Whatever the cause the conditions were ripe for violent action. Although Tom O'Brien and other local officers, as well as Peter Breen and Gabe Dallas from Butte, had continually cautioned the men against any resort to arms, the swelling sense of frustration within the union ranks was nearing the bursting point. The lockout had dragged on for almost half a year; the union's lawyers had failed to set aside the injunctions; and as more scabs were brought in every day the chances of a union victory grew fainter and fainter.

"Up to this time," one observer noted, "the peace party in the union, under the leadership of President O'Brien, have had complete control of the plan of campaign, although not without a great deal of grumbling by the men who thought that severer measures ought to be adopted. The fighting element have sneered at the employment of attorneys, and have said that nothing could come of the fight at Boise; that it was a waste of time and resources, and that the non-union men ought to be run out of the country." They claimed that "the strike was being conducted on milk and water tactics, and that if they had had their way in the beginning the fight would have been won long ago."[24]

Such arguments seemed all the more convincing with the sensational news of July 7th that locked-out steelworkers at Homestead, Pennsylvania, had taken up arms to capture two barge loads of Pinkerton guards and win a bloody but seemingly decisive victory. Just two days later when the miners at Gem discovered that their own union secretary was a Pinkerton spy, they too finally took up arms. Siringo, hiding beneath a boardinghouse floor and crawling along under the boardwalks, eluded the angry miners only increasing their frustration. Tension mounted throughout the following day as armed miners trooped into Gem from Burke and Mullan, and Esler sent in additional arms for the guards and scabs at the Frisco mine and mill.[25]

Early the next morning, July 11th, violence erupted. Shooting began between union miners and the guards barricaded in the Frisco mill at about 5 A.M. The fight soon spread to the Gem mine and sporadic shots continued for several hours before a deafening roar shook the canyon to signal new escalation of the violence. Several miners, breaking into the Frisco powder house, had sent an ore car loaded with dynamite down the incline toward the mill. The fuse was too short and it exploded prematurely, destroying only a portion of the track and the ore bin. Undaunted, the dynamiters moved to the penstock carrying water to the mill and launched a second bundle with a longer fuse. This blast demolished the upper portion of the mill, killing one man and injuring several others in the falling debris. The guards and scabs promptly surrendered. They were disarmed and taken to the miners' union hall. The men at the Gem held out for a few more hours before they too surrendered to end the battle. The captives numbering about 150, were loaded into box-cars and sent down the canyon. In the wake of the battle six men were dead, three on each side, and over a dozen were wounded. But the miners, rejoiced in the thought that they had finally won a clear victory.[26]

Once the fight had begun the miners were determined to make a clean sweep. Late that night more than four hundred armed men headed down the canyon on flatcars and handcars to force the surrender of the three hundred scabs and guards in the Bunker Hill & Sullivan at Wardner. Although their ranks were swelled by nearly two hundred more men at Wardner, they found the mine so well fortified that they decided not to attack. Instead they quietly captured the company's half a million dollar concentrator, and the following morning de-manded that the scabs be discharged within twenty-four hours, or they would blow it up. The manager immediately acceded to the demand. Similar threats were made to the owners of the other mines working scabs. The following morning the last scabs were herded into boxcars and sent out of the Coeur d'Alene.[27]

With the scabs driven out the miners celebrated their victory, seemingly oblivious to the probable outcome of their actions. "No one can realize the feelings of our people," one union man declared. "It has been a long and bitter fight. For months we have been idle and strenuously fighting the grasping encroachments of the mine-owners. It has been a fight for our very existence. We contend that our course is just and we are determined to be recognized. The mine-owners brought scabs in here to kill us if necessary." The miners claimed that the first shot was fired by a Frisco guard who opened fire on a lone union man coming up the canyon, but the owners claimed the miners started the shooting, and all of the subsequent trials never settled the matter.[28]

As far as the miners were concerned the battle was over. The union issued a statement, deploring "the unfortunate affair at Gem and Frisco," and called for a conference with representatives of the MOA. Speaking for the owners, however, A. B. Campbell declared, "All talk of a conference is pure nonsense." For the owners now saw victory within their grasp and vowed to "fight it out to the bitter end."[29]

With the outbreak of violence the MOA was finally able to make a strong case for the imposition of martial law and they promptly called upon both the state and federal governments for troops. This time they found full support in the local press. "Yesterday's violent outbreak of lawlessness in the Coeur d'Alenes was indefensible," the *Spokane Review* charged. "It calls for prompt dispatch of federal troops. . . . It is not necessary to argue about the right or wrong of yesterday's frightful tragedy. When armed bodies of men spurn the law and inaugurate a reign of terrorism, bloodshed and destruction, they forfeit the respect and sympathy of law-abiding people. No amount of specious pleading, no resort to quibbling and hair-splitting can make right such passionate deeds of violence."[30]

In response to the pleas of the mine owners and others, Governor Willey immediately declared a state of martial law in the Coeur d'Alene, calling up six companies of the Idaho National Guard under Adjutant General James F. Curtis to "suppress insurrection and violence." At the same time he sent an urgent plea to President Harrison for federal troops to aid the meager state force of only 191 men. The president responded with a massive force of twenty companies of infantry, swelling the total to nearly 1,500 men. They were placed under the command of Colonel William P. Carlin.[31]

Three companies of black infantry from Fort Missoula, closest to the scene, first tried to enter the Coeur d'Alene on July 12th but they were stopped at a railroad bridge, dynamited the night before. Fearing that the miners still planned to resist, no further attempt was made to move troops into the district until nearly the full force was massed for a single offensive on July 14th. Early that morning the army rolled in by train. They moved cautiously, anticipating more dynamited bridges or possible ambushes. But all was quiet and they met no resistance. Only a handful of curious spectators and a few stray dogs even turned out to meet them. By the end of the day troops were stationed in all of the camps—fourteen companies were bivouacked in and around Wardner and Wardner Junction, where Colonel Carlin made his headquarters, eight at Wallace where General Curtis set up his command, two at Gem, where the trouble had started, and one each at Burke and Mullan. The "insurrection and violence" was thus officially "suppressed" and Curtis telegraphed the governor, "We control the situation."[32]

The following morning the county sheriff and the justices of the peace were all discharged at Curtis's command and the mining company doctor, W. S. Sims, a fiery little southerner "cordially hated by every union man in the county," was appointed sheriff. Curtis also suspended the courts and closed the saloons.[33]

Shortly after noon a heavily armed train, bringing back 250 scabs for the Bunker Hill & Sullivan, rolled into Wardner in what one reporter claimed was "the decisive stroke of the campaign." A squad of "blue coats" were stationed on the front and rear platforms of the scab train with orders to fire at the first sign of danger, and the windows of every car bristled with "long glittering rifle barrels." Although spectators lined the sidewalks, not a word was uttered as the scabs marched up the street through long lines of rifles and bayonets. The union men, who had not expected the scabs to return, were reportedly "crestfallen and sullen."[34]

That same afternoon Curtis posted an order throughout the district, commanding all union members to surrender themselves and their arms. But without waiting for compliance or warrants, squads directed by Sims, Siringo, and others immediately began making "wholesale arrests." One observer predicted "in two days the Coeur d'Alenes will be a vast military prison for the miners' union." Indeed within the next few days more than three hundred men were arrested. They included not only union men but justices of the peace, lawyers, merchants, saloon keepers—anyone who had sympathized with the union and even some who had not. "No man is sure of his liberty," one resident complained, "as it is generally known now that those favoring the cause of the mine owners are pointing out people for arrest to satisfy some personal grudge, or for the enhancement of their own business, and not to promote the ends of justice in connection with the riots."[35]

All of those arrested were herded into two stockades, or "bull pens," at Wallace and Wardner. There they were confined for nearly two months awaiting a hearing to be either charged or released. At first most of the men, considering the arrests a "huge joke," laughed and sang and enjoyed a hearty supper which the ladies of Wallace prepared for them. But they soon began to grumble, as the days and then weeks passed without a hearing and the sanitary conditions grew intolerable in the summer heat. After four weeks the pen at Wardner had to be abandoned for "sanitary reasons" and the prisoners were moved to the pen at Wallace where conditions were only marginally better. Even those on the outside were beginning to complain of "the most noxious odors" pervading the atmosphere.[36]

On July 21st Curtis made an unsuccessful attempt to reduce the number of prisoners by offering parole to all those against whom he felt there was not even sufficient evidence to warrant their being held for examination. But only thirty prisoners, mostly businessmen, took advantage of the offer. All of the rest refused to sign the parole, charging that it was an "undue restraint of their liberty." They demanded instead either to be charged with a crime or released, and their attorneys threatened habeas corpus proceedings if Curtis refused to act.[37]

Twenty-five of the union leaders, including Tom O'Brien and Joe Poynton, were finally charged with contempt of court for violating the federal injunction. On July 25th they were taken to Boise for trial. Three days later, apparently at the insistence of the governor who had begun to worry about the mounting costs, Charles Siringo filed a blanket complaint in federal court against all of the remaining prisoners in order to officially place them under federal jurisdiction and relieve the state of the cost of holding them.[38]

At the same time the Sunday soldiers of the Idaho National Guard were tiring of the adventure and beginning to wail, "Oh, when can we go home." Some, in fact, began to side with the prisoners and one private concluded, "Governor Willey has therefore played a nice hand in favor of the wealthiest men of our state by overlooking the fact that they and their scabs were equally, if not more, to blame than the union men." Even determined patriots, who had convinced themselves that not one of the imprisoned miners was even an "American," griped about the dust and dirt, the lack of blankets and tents, and the poor rations and pauper's pay—43¢ a day! Soon Curtis was barraged with petitions for furloughs and leaves of absence.[39]

Governor Willey was also anxious to end the whole affair, because of its costliness if nothing else. But the MOA, backed by General Curtis, argued that violence would certainly erupt if martial law was not maintained. A compromise was finally reached in late July by recalling five of the six national guard companies, leaving only a token number of state troops in the field. Eight companies of federal troops were also withdrawn soon after, but the twelve remaining companies stayed on throughout the summer and early fall. The returning state militia were assigned to guard the twenty-five union men being brought to Boise for trial. They received a victor's welcome when they arrived in the capital with their captives.[40]

The contempt trial of O'Brien and the other union leaders began on August 2d before U.S. Circuit Judge James H. Beatty,

who had issued the injunction. San Francisco attorney, Patrick
Reddy, who had aided the miners' cause before, joined James
H. Hawley for the defendants against U.S. Attorney Fremont
Wood and MOA attorney Weldon B. Heyburn. Reddy argued
that the MOA had conspired to provoke violence by the importation and hiring of detectives, scabs, and gunmen, and by the
lockouts and injunctions against the union, and that therefore
they were not entitled to redress. But the Judge ruled that provoked or not, dynamiting the Frisco mill and driving out the
scabs were clearly in contempt of court. On August 11th he
sentenced O'Brien and Poynton to eight months in the Ada
County Jail at Boise and gave eleven others terms of four to
six months.[41]

Once the wheels of justice had begun to turn, further indictments and trials followed in quick order. On September 1st
a federal grand jury at Coeur d'Alene City indicted two hundred
union leaders, members, and sympathizers on charges of conspiracy to defeat the progress of justice. Two days later all but
about thirty of those indicted were released on bail or their own
recognizance to await trial; those not indicted were finally given
their liberty; and the bull pen at Wallace was emptied. George
Pettibone and thirteen others were selected for the first conspiracy trial. It got under way on September 9th before a jury,
predominantly of farmers. In answer to the charge of conspiracy, Reddy argued that "concert of action alone does not
constitute conspiracy. The birds sing in concert; but they have
no agreement to do so." But Siringo's inflammatory tales at least
partly convinced the jury and they found four out of the fourteen
guilty. Pettibone was sentenced to two years in the Detroit
House of Correction. The others drew sentences of fifteen to
eighteen months. Reddy immediately appealed the case to the
U.S. Supreme Court and the other conspiracy cases were de-
layed, pending the outcome of the appeal.[42]

In the meantime, however, the state began criminal proceedings and a number of others filed suits of their own. The
Shoshone County grand jury brought in four separate indictments against forty-two of the most prominent union and pro-

union men, including those already convicted of contempt and conspiracy. They were charged with the murders of each of the three guards killed in the battle at Gem and with malicious destruction of the Frisco mill. The jury also indicted Joel Warren for bringing gunmen into the state and Sheriff Cunningham for various incidents of malfeasance during the trouble. At the same time twelve of the scabs injured in the Frisco mill fighting brought suit against A. M. Esler and the Helena & Frisco company for a total of $120,000 in damages. They charged that the company had hired them under the false "representation that there was no trouble, prospective or otherwise, and that their lives and property would be absolutely inviolate." Esler, in turn, sued the county for $100,000 in damages for failure to protect his mill. And when the editor of the *Coeur d'Alene Sun* criticized his action as "an exhibition of imprudence, such as can come only from such an exciter of riots as the black-hearted and lying A. M. Esler," he was slapped with a $15,000 libel suit.[43]

On a change of venue the criminal cases were moved to Kootenai County where the first trial began on November 30th. Webb Leasure, a union sympathizer, was tried for the killing of Ivory Bean, one of the Thiel detectives, guarding the Gem mine. The MOA's attorney, Heyburn, personally prosecuted the case for the state and Siringo provided more sensational testimony, but the jury was not convinced. On December 23d they tendered a verdict of not guilty. The court adjourned for the Christmas holiday and the remaining cases were held over to the March term.[44]

With the establishment of martial law, it seemed to many that the MOA had finally won the decisive round and that the miners' union was completely broken. One enthusiast even ventured that the example set by the MOA would "serve as a guidance in every future strike of the continent." Indeed every union hall in the Coeur d'Alene was locked up; the union's officers and hundreds of members were imprisoned, either awaiting trial or already convicted; hundreds of scabs were working in their places at reduced wages; and nearly all of the MOA's mines and mills were again working to capacity. In August the

MOA had even persuaded General Curtis to close down the Tiger & Poorman mine in Burke, because it was the only major mine still employing union men.[45]

But the union was still very much alive in the minds of its men and even stronger than before in spirit. The long days in the bull pens and jails had only strengthened the determination of most of the men. One confidently predicted,

> it is only a matter of time when the union will be as strong as ever. The mine owners have the best of it now, but they will resume their arrogant demands again and introduce the company boardinghouse and store. This will arouse the scabs now working, and they will begin to organize to protect themselves. They will see the justice of the present strike and they will be as bitter enemies of the association as ever we were . . . and once again the old struggle will come up. . . . In the end we must win and we are determined to do so if it takes ten years.[46]

Governor Willey finally declared an end to the five-month reign of martial law after the elections in November. As soon as the last troops were withdrawn, the union slowly began to rebuild. Although most of the officers were still in jail, regular union meetings were resumed. The owners claimed not to discriminate against union men, but any whom they suspected of having taken part in the violence found themselves blacklisted. Still, many union men again found work in the mines. One by one they began winning over the scabs to their cause, and a mass conversion of scabs occurred the following year when the Bunker Hill & Sullivan and other MOA mines closed down because of declining lead prices.[47]

The outcome of the court trials also gave much "satisfaction" to many of the union men. They viewed the failure of the state to convict Webb Leasure as a triumph for the union, since the MOA had openly boasted that they would hang him, as their strongest case was against him. As one miner noted, "Leasure's trial places the mine owners in a very bad light, and puts their case before the people in the light of a persecution, instead of a prosecution." Indeed when the court resumed in March, the district attorney asked that all of the state's other cases against the miners be quashed, conceding that he had no evidence to warrant conviction in any of the cases.[48]

The final "vindication" of the miners came on March 6th, when the U.S. Supreme Court dismissed the federal conspiracy indictments on a technicality and reversed the convictions, freeing Pettibone and others from the penitentiary. The Butte labor paper, the *Bystander*, hailed the decision as "a body blow to the Miner Owners' Association and its subsidized courts." As the decision also cast doubt on the contempt indictment, Judge Beatty quashed that himself, freeing the last union men from the Ada County Jail.[49]

Early in April Tom O'Brien and Joe Poynton returned to the Coeur d'Alene to a hero's welcome. At each camp they were greeted by brass bands and cheering crowds, and honored with banquets and speeches. "No men," the *Mullan News* reported, "have ever been so honored before in the Coeur d'Alenes." O'Brien, opposed to violence from the beginning, took the opportunity to again call for peace, "We think it is now time, that the differences existing between ourselves and the mine owners should be amicably adjusted," and he looked forward to a time when "we can settle down and work in harmony with our employers."[50]

But the long battle in the Coeur d'Alene had settled nothing; if anything, it had only embittered many of the participants. The issues of dispute remained unchanged. The MOA was still determined to break up the union and the union still demanded a $3.50-a-day minimum wage for all men underground. Declining lead prices triggered a new crisis. On March 1, 1893, the Bunker Hill & Sullivan closed and others followed within the next few months. By the first anniversary of the battle at Gem all of the major mines in the district had closed again. The owners again talked of resuming work only after new wage cuts; the miners again vowed to "give the mine owners a fight to the finish"; and the battle lines were sharply drawn once more. This first battle in the Coeur d'Alene had been but the opening engagement in a far more devastating war that would rage on for over a decade with ever escalating violence and on ever widening fronts until it engulfed nearly all the western mines in a turbulent and bloody conflict.[51]

Already the war had spread far beyond the Coeur d'Alene. The failure of the Coeur d'Alene Mine Owners' Association to break up the union and lower wages was not yet apparent, when mine owners elsewhere in the West adopted a similar aggressive course of lockouts and wage cuts. Spurred by declining lead prices and enticed by the seeming victory of the owners in the Coeur d'Alene, they launched a concerted attack on wages and the unions in the winter of 1892–93.

The Utah mine owners, who were already paying miners only $3 a day, led this attack. On New Year's Day 1893 the owners of the principal mines in Bingham Canyon discharged all their men, announcing they would not resume at more than $2.50 a day. Within the next few weeks owners at Eureka and Mammoth in the Tintic district joined the attack, and more than a thousand men were thrown out to tramp the Utah snows. Within a month the assault spread through Colorado and Montana, as mine owners at Aspen, Castle, Granite, Neihart, Red Mountain, Rico, Summitville, and Telluride, all closed down, demanding wage cuts to $3 a day, and a few thousand more men were locked out in the snow. The spirit of the attack even carried into that bastion of the $4 wage, the Comstock, where a committee of superintendents announced that they would ask the unions to again consider a graduated wage scale.[52]

The union miners opposed the wage cuts, but to some observers their cause seemed foredoomed. "A tremendous effort is being made everywhere to check the inevitable fall in wages," one mining editor wrote. "But it cannot work . . . no power on earth except an advance in silver can keep them up." With thousands of men out of work the supply of labor exceeded the demand, he argued, so strikes against the wage cut had "as little chance of success as a strike against the laws of God."[53]

Despite such doomsayers the miners were determined to resist the cuts. But their only recourse against the lockout was either to try to outwait the owners or to try to negotiate a compromise. In February the Aspen union won a speedy settlement by agreeing to work longer hours at the old wage. Not all attempts to negotiate were so successful, however. Early in March the Eureka miners tried to break the deadlock there by proposing a sliding scale tied to bullion prices, but H. E. Hyde of the Bullion-Beck mine mistakenly saw this offer as a sign that the union was about to give up the fight. He rejected the compromise and brought in scabs. He was forced to turn back with his first trainload of scabs, when the miners greased the rails on the steep grade up to the camp. But protected by armed guards and detectives, the following week he succeeded in bringing in a hastily gathered gang of sheepherders and Mormon farmhands. Even then the boardinghouse hashers joined with the miners to persuade the bulk of them to quit. Tensions slowly mounted. Several scabs and miners came to blows. Hyde's brother, Hyrum, shot a miner in a heated argument and was shot in turn by another; both were only wounded. Finally a dynamite charge was set off behind a cabin where scabs were living. Hyde and other owners denounced the outrage, calling upon the governor for protection. But he declined, so the dispute dragged on.[54]

Most of the unions were not financially prepared to sustain a long fight, and with so concerted an assault by the owners even the Butte and Black Hills miners, still sending aid to the Coeur d'Alene, could not provide support on so many new fronts. In February the Rico union conceded defeat, accepting the wage cut in exchange only for the privilege of boarding wherever they chose. As other unions began to falter and other mine owners began to talk of wage cuts, it became clear that the very existence of the mining labor movement would soon be threatened. Thus many of the unions throughout the West eagerly turned to some form of federation to build stronger bonds between them and aid them and unite them in their struggle for survival.[55]

The beleaguered unions in the silver-lead camps urgently sought the moral and financial support a federation could provide. Although the unions in the copper and gold camps, such as Butte and the Black Hills, were not immediately threatened, they had funded the fights against wage cuts elsewhere to check the movement before it became a direct threat to them. And as this had heavily taxed their own resources, they too strongly favored a federation to broaden the financial base for support of the struggle.

The idea of uniting all of the miners' unions in the West had been discussed for some time, so when the decision to act was finally made by Tom O'Brien and Joe Poynton in the Ada County Jail in March of 1893, the work was quickly completed. The Butte union issued the call and within weeks the Western Federation of Miners was born. The concerted assault by the owners had finally united all the miners. And as Ed Boyce, president of the Federation later reflected, the Coeur d'Alene had been the spark: "It may be truly said that the miners of the West made no attempt at forming a central body until the Mine Owners' Association, under the guiding hand of John Hays Hammond and his associates, forced the issue upon them in the Coeur d'Alenes in 1892." Although the mining war in the West would rage on for over a decade, the miners were now solidly organized and equal to the fight.[56]

The Federation

In May of 1893, as the labor war gripped the western states, the hardrock miners gathered in the Miners' Union Hall at Butte to form "a grand federation of underground workers throughout the western states"—the Western Federation of Miners. This federation would not only rally the existing unions for their own survival, but would go on to organize nearly all the workers in the industry, both below and above ground, becoming the strongest labor organization in the West. In the militant tradition of the earlier unions, it would also become "the most militant in the history of the United States," and the power behind those arch radicals, the Industrial Workers of the World.[1]

The mine owners, who had fought so bitterly against even the independent unions, would become nearly rabid in their fight against the federation—to them the "Western Federation of Dynamiters and Murderers." "The Federation," they would rave, "is an outlaw with the avowed purpose of destroying existing government," and they would see their fight as a glorious crusade "for personal and business freedom, . . . for freedom of speech and the press, *for the protection of 'the present system of government.'*" Indeed, in their zeal they would soon declare "almost any means are justifiable to rid the mining community of the anarchists who lead the Western Federation of Miners."[2]

But many other observers would find much to praise in the federation. "In its attitude toward the working class," one prominent journalist would write,

the Western Federation has displayed an idealism which has brought a ray of imagination and of sentiment into the life of many an underground toiler. Opening its doors freely and gladly to all workingmen, denouncing all devices for excluding outsiders and for making the trade union a monopoly, cherishing the interests of the unskilled man even above those of his more fortunate comrade, preaching the doctrine of a united working class, calling upon every workingman to regard his brother's trials and ambitions as his own, fighting success-

fully for the establishment of eight-hour laws, offering to the anarchism of certain corporations the only real resistance which that anarchism has ever encountered, the Western Federation of Miners has
contributed to the history of Western mining its one flash of social
thought, its one deviation from a materialistic line of progress.[3]

More than forty delegates convened at Butte on May 15, 1893,
to form the Western Federation of Miners. They came from the
unions at Aspen, Creede, Ouray, and Rico in Colorado, from
Burke, Gem, and Mullan in Idaho, from Bannack, Barker, Belt
Mountain, Butte, and Granite in Montana, from Central City,
Lead City, and Terry's Peak in South Dakota, and from Eureka
and Mammoth in Utah. Even smaller unions, that were unable
to send delegates, sent word that they too were "heartily in
sympathy with the federation and will fall in line at the first
opportunity." The only notable absences were the Comstock
and Bodie unions. The reason for their absence is not clear, but
it seems likely that it was talk of establishing "one scale of wages
for all" that kept them away. For it was obvious that if a single
scale were adopted, it would be something less than the $4 a
day which they were still fighting to maintain. The miners' assemblies of the Knights of Labor in Washington, Oregon, and
parts of Colorado, on the other hand, were simply not invited
because they had agreed to wage scales that were too low.[4]

Even if the Comstock unions were not directly represented,
however, a number of former Comstock miners were among
the delegates. The most prominent were Thomas O'Brien and
Joseph F. Poynton, who had initiated the move to federate, and
old Patrick Henry Burke, who had been active in the mining
labor movement throughout most of its history, since the founding of the Miners' League of Storey County in 1864. As the
number of delegates from each union was proportional to its
membership, the Butte delegation was by far the largest, making up forty percent of the total, and with their Montana
"branches" they held a solid majority. The Butte delegates were
led by their president John L. Williams, and they included such
veteran union men as James H. Rowe, former president of the
Ruby Hill Miners' Union, Charles O'Brien, former president of
the Broadford union, and Thomas M. Malouin, one of the found-

ers of the Tuscarora union—all of whom had come to Butte after the breakup of their unions in the $4 fight.[5]

The delegates generally were men well seasoned in the mining labor struggles of three decades. They and the other hardrock miners they represented knew the value of organization and solidarity, and they needed no convincing of the merits of federation. As the first order of business on May 15th they unanimously voted to form the Western Federation of Miners. A committee with one man from each union was appointed to draft a constitution and bylaws. In three days their task was completed.[6]

"The object of this Federation," they set forth in the constitution, "is to unite the various Miners' Unions of the West into one central body; to practice those virtues that adorn society and remind man of his duty to his fellow-men; the elevation of his position and the maintenance of the rights of the miner." Here too the ubiquitous influence of the Comstock unions was clearly evident. The second objective was taken directly from the constitution of the Virginia City Miners' Union and the third from that of the Gold Hill Miners' Union.[7]

The specific goals of the federation were outlined in the Preamble:

Since there is scarcely any fact better known than that civilization has for centuries progressed in proportion to the production and utilization of the metals, precious and base, and most of the comforts enjoyed by the great majority of mankind are due to this progress, the men engaged in the hazardous and unhealthy occupation of mining should receive a fair compensation for their labor, and such protection from the law as will remove needless risk to life and health; we therefore deem it necessary to organize the Western Federation of Miners of America for the purpose of securing by education and organization and wise legislation a just compensation for our labor and the right to use our earnings free from dictation by any person whatsoever. We therefore declare our objects to be:

FIRST: To secure an earning fully compatible with the dangers of our employment.

SECOND: To establish as speedily as possible and forever our right to receive pay for labor performed in lawful money, and to rid ourselves of the iniquitous system of spending our earnings where and how our employers or their officers may designate.

THIRD: To procure the introduction and use of any and all suitable, efficient appliances for the preservation of life, health and limbs of all employees, and thereby preserve to society the lives of large numbers of wealth producers annually.

FOURTH: To labor for the enactment of suitable mining laws, with a sufficient number of inspectors, who shall be practical miners, for the proper enforcement of such laws.

FIFTH: To provide for the education of our children by lawfully prohibiting their employment until they shall have obtained a satisfactory education, and in every case until they shall have reached sixteen years of age.

SIXTH: To prevent by law any mine owner or mining company from employing any Pinkerton detectives or other armed forces for taking possession of any mine, except the lawfully elected or appointed forces of the state, who shall be bona fide citizens of the county and state.

SEVENTH: To use all honorable means to maintain friendly relations between ourselves and our employers, and endeavor by arbitration and conciliation to settle such differences as may arise between us, and thus make strikes unnecessary.

EIGHTH: To use all lawful and honorable means to abolish the system of contract convict labor in the states where it now exists and to demand the enforcement of the foreign contract labor law and protection of our American miners and mechanics against imported pauper labor.

NINTH: To demand the repeal of all conspiracy laws that in any way abridge the rights of labor organizations.

TENTH: To procure employment for our members in preference to non-union men.[8]

The only important issue not mentioned here was the fundamental one of whether or not to set a uniform minimum wage throughout the West. This was "the hardest problem with which the convention has to wrestle," one observer noted, since different wages and hours prevailed in most of the camps represented. After much discussion the delegates finally decided "that each union in the federation should settle the question of wages to be received within their jurisdictions." This cleared the way for the Comstock and Bodie unions to join the federation without compromising their position on the $4 wage.[9]

The first officers of the Western Federation were president John Gilligan of Butte, first vice-president D. D. Goode of Granite, second vice-president John Duggan of Eureka, secretary-treasurer Thomas M. Malouin of Butte, warden William

Cunningham of Butte, and an executive committee of five: John McLeod of Terry's Peak, James Millett of Granite, and Anthony Mathews, Patrick Gallagher, and John Gilligan, all of Butte. Between conventions the full power to direct the Federation was in the hands of the Executive Committee. As only a simple majority was required, the Butte miners directly held this power. This included disbursement of relief funds to locked-out or striking unions and the levying of assessments on the locals to support such relief. The federation also maintained a general fund, supported by an annual levy of one dollar on every miner in good standing, to be paid from the general fund of each local union.[10]

One of the most important provisions in the federation's constitution was that allowing the appointment of "organizers ... who shall diligently labor to organize all non-union miners." They were to be paid $3 a day plus expenses. In the past union organization in the western camps had frequently been spurred only by crisis and often too late to be effective. Only the Comstock and Butte unions had made any persistent effort to encourage organization in the neighboring camps or send officers to help set up a union once a movement had begun. But the need for a strong effort at organization was obvious. There were more than thirty thousand hardrock miners in the West by that time and barely a quarter of them—only those in the larger camps—were organized. The mine owners bitterly denounced these new organizers, but the rapid growth of the federation and its resultant strength over the next decade was attributable in large measure to their "diligent" work.[11]

In general the constitution adopted for the federation dealt with matters relating to the body as a whole and it only supplemented the existing constitutions of the local unions. These were all derived from those of the Comstock unions and they continued in effect for all local matters under the federation. Only one provision of the federation constitution directly affected local action. This specified, "It shall be unlawful for any local union to enter upon a strike unless when ordered by three-fourths of its resident members and on approval of the Executive Board." Previously some unions had struck, when the decision was supported only by a simple majority of those *voting*, and such strikes often failed for want of support. This provision, requiring a much stronger mandate, helped ensure the solidarity necessary for success.[12]

On the fifth and last day of the convention the delegates adopted a lengthy series of resolutions, dealing with issues of particular interest to western miners and a few of broader national interest. They called for "unlimited coinage" of silver and gold, for repeal of all conspiracy and antiboycott laws, for prohibition against the use of "Pinkerton and other so-called detective armed forces," for disbanding of state militia companies, for more strenuous laws against foreign contract labor, and for enactment of eight-hour laws and mining safety laws, enforced by state mining inspectors. They also supported popular election of United States senators, construction of the Nicaraguan canal, and nationalization of the railroads and telegraph. Most of the resolutions were warmly endorsed by the local press and the convention adjourned.[13]

Thus the Western Federation of Miners was born. Within the next few weeks fourteen of the seventeen unions represented at the convention applied for charters to become locals. The Butte Miners' Union was named Local No. 1 and the offices of the federation were set up in their hall. Only the Bannack, Creede, and Mammoth unions—which collapsed that summer—failed to join.[14]

Although the western miners were now more solidly united than ever before, their outlook for the future was still bleak. In little over a month after the founding of the federation, the silver and lead mining industry was plunged into deeper financial trouble. The closing of the Indians mints, the repeal of the Sherman Silver Purchase Act, and the general panic of 1893 sent silver and lead prices crashing to new lows. Silver fell from 84¢ an ounce to barely 60¢ and lead from over 4¢ a pound to nearly 3¢. Many more mines throughout the West were forced to close and the movement to cut wages became rampant.[15]

The federation took over the relief for the locked-out Coeur d'Alene and Eureka miners. But despite this aid the Eureka union finally capitulated in August. Their wage cut from $3 to

$2.50 a day was only partly offset by a reduction in the cost of board at the company boardinghouses. The Coeur d'Alene miners held out, but wage cuts followed in most of the other silver-lead camps. In September the Leadville miners' assembly of the Knights of Labor agreed to a scaled cut to $2.50 a day as long as silver remained below 83½¢ an ounce. The Aspen miners' union settled for a more elaborate scale the following month, setting wages at $2.25 a day when the price of silver was between 70¢ and 75¢ an ounce, $2.50 between 75¢ and 80¢, $2.75 between 80¢ and 85¢, and $3 between 85¢ and 90¢. Even the Comstock miners were threatened with a wage cut. But, bolstered by promises of support from the Butte union, they rejected a proposed cut by a vote of 229 to 187 and the owners chose not to press the matter to the point of a strike.[16]

Silver had been the principal product of the western mines since the 1870s and the silver camps had been the mainstay of the mining labor movement, but the collapse of silver prices in 1893 brought this to an end for all time. Although lead prices soon recovered to sustain some of the largest silver-lead mines, silver prices never recovered to more than 70¢ an ounce during the next quarter century and most of the silver mines were closed permanently. What saved the western mining industry, and the mining labor movement, was the revival of gold mining and the steady growth of copper mining. The decline in silver prices sent prospectors back into the hills looking for gold again and the discovery of fabulously rich lodes at Cripple Creek sparked a new excitement that swept the West. Capital slowly shifted to the new gold and copper camps and within a few years the mining industry was prosperous once again.

With determined, aggressive leadership the Western Federation of Miners also began to grow, but since the mine owners fought it at every step, the mining war raged on relentlessly. The course of the federation was faltering at first, handicapped by the chronic Irish-Cornish wrangling within the Butte union. Several of the federation officers from Butte resigned precipiously, leaving the federation to drift almost leaderless, and with little attempt yet made at organizing, there were fewer unions represented at the second annual convention than at the first.

But in 1896 Butte control over the federation was finally broken with the election of Ed Boyce of the Wardner union as president. Boyce, radicalized in the bullpens of the Coeur d'Alene, gave the federation the enthusiastic and aggressive leadership it needed. Under his direction it grew to more than two hundred locals within the next five years to clearly become the most powerful labor union in the West and the most militant champion of industrial unionism in the nation.

Indeed, the tradition of militant industrial unionism, fostered by the early miners' unions, reached its fullest expression in the Western Federation of Miners, and ultimately sparked the radical socialism of the Industrial Workers of the World. From the first labor agitation on the Comstock, the miners had rejected "pure and simple" trade unionism to embrace industrial unionism. Vowing to "boldly defy . . . the tyrannical, oppressive power of Capital," the Comstock unions had lead the way in organizing all underground workers and demanding a uniform minimum wage for all, regardless of their level of skill. This policy was sharply attacked by trade unionists as well as by the mine owners, but it remained a fundamental principle of the western mining labor movement. In defense the miners argued that all faced equally the risk of injury or death underground and that this risk, not their various skills, should be the true basis for their wage. The early unions fought for this point nearly every time they organized and they always won it, if they survived at all. Occasionally they flirted with even broader unionism, organizing the mills too, or even the entire camp, as the early Leagues and Workingmen's Unions testify.

The Western Federation's militant pursuit of industrial unionism was thus the irrepressible consequence of the traditions and struggles of three decades and of the drive of the leaders raised in them. The changing nature of the western mining industry after the collapse of silver in 1893 only accelerated the pursuit. The increasing development of lower-grade, base-metal deposits led inevitably to more massive capitalization and larger, often open pit, operations employing more mechanized equipment and less-skilled workers. This further widened the breach between the workers and the owners, further blurred the distinctions between the skilled and the unskilled, and even between the underground and the surface. With increasing militancy the federation soon embraced not only all mine workers, but all mill and smelter workers as well. The subsequent history of the Western Federation of Miners was even more turbulent than that of the earlier unions that formed it, but this story has already been forcefully told by Vernon H. Jensen in his *Heritage of Conflict.*

Looking back over the years from 1863 to 1893, we have seen the emergence of a militant labor movement in the western metal mines. It was the necessary response to the rapid industrialization of the mines which stifled the feelings of freedom and optimism that characterized the surrounding frontier. It was committed to militant industrial unionism from its inception and it set the course for labor relations in the industry for years to come. Although wages were most often the center of its disputes with management, the bulk of its resources and energies were devoted to aiding the sick and bereaved. But as it grew its goals became increasingly varied and complex. Its program expanded from the simple maintenance of a minimum wage and sick benefits to demands for a closed shop, an eight-hour day, safety legislation, and improved working and living conditions. Its courses of action broadened to include not only strikes but boycotts, political action, and legislative redress. But with ever increasing industrialization came ever increasing violence.

It might seem easy to attribute this violence in the western mines to the lawlessness of the frontier, as a number of writers, have, but this is not the case. In the early disputes, when law enforcement was least effective, the miners acted with the greatest restraint. Only as the law grew stronger and the owners began to manipulate it as a tool of repression, and only as the freedom of the frontier faded, did the miners in their frustration turn to violence. Both the labor repression and violence in the western mines were, in fact, but an imitation of that already rampant in the more settled and civilized eastern states. Troops and hired gunmen were called in to break strikes in the coal pits of the East well before they were used in the Amador war and in Leadville; the same may be said of blacklisting, Pinkerton spies, lockouts, and injunctions; even the Haymarket dynamiting preceded the dynamiting at Grass Valley where it had been peacefully used for nearly two decades; and it was the violent eruption at Homestead that helped to trigger that in the Coeur d'Alene. Thus it was the "taming of the frontier," not frontier lawlessness, that spurred the violence—that heritage of conflict—that left so lasting a scar on labor relations in the western mines.

Notes

Notes

THE HARDROCK MINER

1. Rossiter W. Raymond, *Statistics of Mines and Mining . . . for the Year 1870* (Washington: G.P.O., 1872), p. 3.

2. San Diego, *Union and Bee*, March 1, 1889.

3. An excellent history of the development of western mining is given in Rodman W. Paul, *Mining Frontiers of the Far West 1848–1880* (New York: Holt, Rinehart and Winston, 1963). Lode mining statistics are summarized in the *Ninth Census, 1870, III* (Washington: G.P.O., 1872), pp. 767–780; *Tenth Census, 1880, XIII* (Washington: G.P.O., 1885), pp. 156–157; *Eleventh Census, 1890, VII* (Washington: G.P.O., 1892), pp. 33 ff.

4. Silver City, Idaho, *Owyhee Avalanche*, April 22, 1871.

5. Raymond, *op. cit.*, p. 4.

6. Richard H. Morefield, "Mexicans in the California Mines, 1848–1853," *California Historical Society Quarterly*, 35 (1956), 37–46; William Wright (Dan De Quille), *History of the Big Bonanza* (Hartford, Conn.: American Publishing Co., 1876), p. 85; Eliot Lord, *Comstock Mining and Miners* (Washington: G.P.O., 1883), p. 386; Richard J. Hinton, *The Handbook to Arizona* (San Francisco: Payot, Upham and Co., 1878), pp. 213, 267–268.

7. Silver City, Idaho, *Owyhee Avalanche*, April 22, 1871; Arthur C. Todd, *The Cornish Miner in America* (Truro: D. Bradford, 1967); Lynn I. Perrigo, "Cornish Miners of Early Gilpin County," *Colorado Magazine*, 14 (1937), 92–101; Myron Angel, ed., *History of Nevada* (Oakland: Thompson and West, 1881), p. 657.

8. Silver City, Idaho, *Owyhee Avalanche*, April 22, 1871.

9. Lord, *op. cit.*, pp. 382–386; Wright, *op. cit.*, p. 328.

10. Lord, *op. cit.*, pp. 383–384; Charles H. Shinn, *The Story of the Mine, as Illustrated by the Great Comstock Lode of Nevada* (New York: D. Appleton and Co., 1896), p. 242.

11. Lord, *op. cit.*, pp. 379–380.

12. *Ibid.* p. 372; Albert Williams, "Modern Types of Gold and Silver Miners," *Engineering Magazine*, 2 (Oct. 1891), 48–62.

13. San Francisco, *Mining and Scientific Press*, March 8, 1890.

14. *First Biennial Report of the Bureau of Labor Statistics of the State of Colorado, 1887–1888* (Denver, 1888), pp. 250–253; Lord, *op. cit.*, pp. 372–373, 381; Eureka, Nev., *Daily Sentinel*, Jan. 18, 1877.

15. *Ibid.*

16. Lester A. Hubbard, *Ballads and Songs from Utah* (Salt Lake City: University of Utah Press, 1961), p. 435.

17. Eureka, Nev., *Daily Sentinel*, Jan. 18, 1877.

18. Lord, *op. cit.*, pp. 377–378; Williams, *op. cit.*, p. 59; A. Burrows, "Social Life among Miners," *Sunset*, 16 (March 1906), 434–435.

19. Candelaria, Nev., *True Fissure*, March 11, 1882.

20. Shinn, *op. cit.*, p. 225.

21. Candelaria, Nev., *True Fissure*, March 11, 1882.

22. Lord, *op. cit.*, pp. 389–390.

23. *Ibid.*, pp. 393–395.
24. *Ibid.*, pp. 397–398.
25. *Ibid.*, p. 374.
26. *Ibid.*, p. 401.
27. Rossiter W. Raymond, "The Hygiene of Miners," *Transactions of the American Institute of Mining Engineers,* 8 (1879), 108–109.
28. Jewett V. Reed and A. K. Harcourt, *The Essentials of Occupational Diseases* (Springfield, Ill.: Charles C. Thomas, 1941), pp. 20–31.
29. *Ibid.*, p. 76; Raymond, "The Hygiene of Miners," p. 113.
30. Reed and Harcourt, *op. cit.*, pp. 50–64; *Tenth Census, 1880*, III, pp. 175, 438.
31. Reed and Harcourt, *op. cit.*, pp. 161–174; San Francisco, *Mining and Scientific Press,* Feb. 18, 1893; *Final Report and Testimony Submitted to Congress by the Commission on Industrial Relations, 64th Congress, 1st Session, Senate Document 415* (Washington: G.P.O., 1916), IV, 3942–3943, 3949, 3960–3965.
32. Williams, *op. cit.*, p. 48.
33. Harry M. Hanson, "Gold Mining in the Black Hills," *Engineering Magazine,* 3 (1892), 691; Joseph Harper Cash, "Labor in the West: The Homestake Mining Company and Its Workers, 1877–1942," Unpublished Ph.D. Dissertation, University of Iowa, 1966, p. 95.
34. Hanson, *op. cit.*, pp. 691–693; William P. Blake, *Notices of Mining Machinery and Various Mechanical Appliances in Use Chiefly in the Pacific States and Territories for Mining, Raising and Working Ores* (New Haven: Charles Chatfield and Co., 1871), p. 14.
35. W. L. Saunders, "History of Rock Drills," *Mining and Scientific Press,* May 21, 1910; Grant H. Smith, *The History of the Comstock Lode 1850–1920,* University of Nevada Bulletin, vol. 37, no. 3 (Reno, 1943), p. 246; Blake, *op. cit.*, pp. 33–42.
36. Hanson, *op. cit.*, pp. 693–695; Blake, *op. cit.*, p. 19.
37. Hanson, *op. cit.*, p. 695; Cash, *op. cit.*, pp. 97–98.
38. Joseph S. Curtis, *Silver-Lead Deposits of Eureka, Nevada* (Washington: G.P.O., 1884), p. 150; Wright, *op. cit.*, pp. 427–428.
39. Eureka, Nev., *Daily Sentinel,* March 7, 1876.
40. Curtis, *op. cit.*, p. 151; Ruby Hill, Nev., *Mining News,* March 19, July 28, 1884; Eureka, Nev., *Daily Sentinel,* June 23, Oct. 20, 22, 1885, June 22, 1886.
41. San Francisco, *Mining and Scientific Press,* Feb. 17, 1906.
42. John A. Church, "Accidents in the Comstock Mines and Their Relation to Deep Mining," *Transactions of the American Institute of Mining Engineers,* 8 (1879), 84–97; Lord, *op. cit.*, p. 404; *Tenth Census, 1880,* XIII, pp. 173–177.
43. *Constitution, By-Laws, Order of Business and Rules of Order of the Miners' Union of Gold Hill, Nev.* (Virginia City: Enterprise, 1871), p. 3; San Francisco, *Mining and Scientific Press,* Feb. 12, 1881.
44. Lord, *op. cit.*, pp. 399, 404; Wright, *op. cit.*, p. 202.
45. *Ibid.*, pp. 211, 213.
46. Lord, *op. cit.*, pp. 398–399.
47. *Ibid.*, p. 401.
48. *Ibid.*, pp. 219–220.
49. San Francisco, *Mining and Scientific Press,* Feb. 20, 1904.
50. Church, *op. cit.*, pp. 92–93.
51. Wright, *op. cit.*, p. 176.
52. *Ibid.*, pp. 177–183.
53. Lord, *op. cit.*, pp. 404–405.
54. *Constitution . . . of the Miners' Union of Gold Hill,* p. 3.

55. London, *Mining Journal*, May 7, 1881, quoted in Clark C. Spence, *British Investment and the American Mining Frontier 1860–1901* (Ithaca, N.Y.: Cornell University Press, 1958), pp. 98–99. Spence discusses the management problems of British companies in the West.

56. Gold Hill, Nev., *Daily News*, Aug. 3, 9, 1877.

57. Grass Valley, Calif., *Foothill Tidings*, Feb. 17, 1877; Hailey, Idaho, *Wood River Times*, April 9, 1884, Sept. 1, 1885, May 1, 1889.

58. New York, *Engineering and Mining Journal*, July 5, 1902.

59. San Francisco, *Mining and Scientific Press*, May 28, 1881; Los Angeles, *Daily Herald*, May 25, 1881; Independence, Calif., *Inyo Independent*, May 28, 1881.

60. Eureka, Nev., *Daily Sentinel*, Feb. 24, March 2, 1876; Austin, Nev., *Reese River Reveille*, May 28, 1875, March 30, 1881; "Constitution and By-Laws, Pledge and Minutes of the Miners' Union of the Town of Gold Hill, State of Nevada," p. 31. Manuscript in Special Collections of the University of Nevada Library, Reno, Nevada.

THE COMSTOCK UNIONS

1. Rodman W. Paul, *Mining Frontiers of the Far West 1848–1880* (New York: Holt, Rinehart and Winston, 1963), pp. 56 ff., describes the technological development of deep mining on the Comstock.

2. Virginia City, *Territorial Enterprise*, May 31, 1863, quoted in Eliot Lord, *Comstock Mining and Miners* (Washington: G.P.O., 1883), p. 183; Myron Angel, ed., *History of Nevada* (Oakland, Calif.: Thompson and West, 1881), p. 261.

3. Hubert H. Bancroft, *History of Nevada, Colorado and Wyoming, 1540–1888* (San Francisco: The History Co., 1890), p. 130.

4. Gold Hill, *Daily News*, March 21, 1864; Virginia City, *Territorial Enterprise*, March 22, 1864, quoted in Lord, *op. cit.*, pp. 183–184.

5. Gold Hill, *Daily News*, Aug. 1, 1864; Virginia City, *Daily Union*, Aug. 2, 1864; San Francisco, *Mining and Scientific Press*, Aug. 6, 1864; George R. Brown, ed., *Reminiscences of Senator William M. Stewart of Nevada* (New York and Washington: Neale Publishing Co., 1908), pp. 164–165.

6. Gold Hill, *Daily News*, Aug. 1, 1864; Virginia City, *Daily Union*, Aug. 2, 1864; Lord, *op. cit.*, p. 185.

7. Gold Hill, *Daily News*, Aug. 2, 1864.

8. Gold Hill, *Daily News*, Aug. 1, 1864; Virginia City, *Daily Union*, Aug. 2, 1864; Virginia City, *Territorial Enterprise*, Aug. 16, 1864, quoted in Lord, *op. cit.*, p. 187.

9. Gold Hill, *Daily News*, Aug. 3, 1864; Virginia City, *Daily Union*, Aug. 2, 3, 1864; Angel, *op. cit.*, p. 605.

10. Gold Hill, *Daily News*, Aug. 8, 1864; Virginia City, *Daily Union*, Aug. 7, 1864.

11. Gold Hill, *Daily News*, Aug. 8, 1864.

12. *Ibid.*, Sept. 16, 1864.

13. *Ibid.*, Aug. 8, 1864; Virginia City, *Daily Union*, Aug. 7, 1864.

14. Gold Hill, *Daily News*, Aug. 29, Sept. 1, 16, 1864.

15. Aurora, Nev., *Esmeralda Union*, Aug. 1, 2, 1864, quoted in Virginia City, *Daily Union*, Aug. 7, 1864; Sacramento, *Daily Union*, Aug. 4, 1864; San Francisco, *Mining and Scientific Press*, Aug. 13, 1864; Austin, Nev., *Reese River Reveille*, Sept. 16, Oct. 6, 1864; Virginia City, *Daily Union*, Aug. 9, Sept. 20, 1864.

16. Lord, *op. cit.*, p. 186.

17. Gold Hill, *Daily News*, Sept. 16, 29, 1864.

18. *Ibid.*

19. *Ibid.*, Sept. 21, 1864; Virginia City, *Daily Union*, Sept. 23, 1864.

20. Gold Hill, *Daily News*, Sept. 23, 1864.

21. *Ibid.*

22. *Ibid.*, Sept. 24, 1864.

23. *Ibid.*, Sept. 20, 29, 1864; Sacramento, *Daily Union*, Oct. 3, 7, 1864.

24. Virginia City, *Territorial Enterprise*, Aug. 25, 1865, cited in Lord. *op. cit.*, p. 190.

25. "Constitution and By-Laws, Pledge and Minutes of the Miners' Union of the Town of Gold Hill, State of Nevada," p. 41. Manuscript in Special Collections of the University of Nevada Library, Reno. Biographical sketches of John G. White are given in Harry L Wells, *History of Nevada County, California* (Oakland: Thompson and West, 1880), p. 233; *A Memorial and Biographical History of Northern California* (Chicago: Lewis Publishing Co., 1891), pp. 367–368.

26. "Constitution . . . Gold Hill, State of Nevada," p. 1.

27. *Constitution, By-Laws, Order of Business and Rules of Order of the Miners' Union of Gold Hill, Nev.* (Virginia City: Enterprise, 1871), p. 3.

28. "Constitution . . . Gold Hill, State of Nevada," pp. 1, 31.

29. Lord, *op. cit.*, pp. 383–386.

30. *Constitution and By-Laws of the Washoe Typographical Union including the scale of prices and list of members. Organized June 28, 1863* (Virginia, Nevada Terr.: Standard Book and Job Printing Office, 1863); A. K. H. Jenkins, *The Cornish Miner: An Account of His Life Above and Underground from Early Times* (London: George Allen and Unwin Ltd., 1927), p. 332; Sidney Webb and Beatrice Webb, *The History of Trade Unionism* (New York: Longmans, Green and Co., 1920), p. 434; D. B. Barton, *Essays in Cornish Mining History* (Truro: D. Bradford Barton, 1968), pp. 46 ff.; Edward A. Wieck, *The American Miners' Association: A Record of the Origin of Coal Miners' Unions in the United States* (New York: Russell Sage Foundation, 1940), pp. 9, 86, 93.

31. "Constitution . . . Gold Hill, State of Nevada, pp. 41–53.

32. *Ibid.*, pp. 58–68, 83; Gold Hill, *Daily News*, Feb. 12, 1867.

33. "Constitution . . . Gold Hill, State of Nevada," pp. 63, 70; Gold Hill, *Daily News*, Feb. 11, 12, 1867; Virginia City, *Daily Trespass*, Feb. 12, 1867.

34. "Constitution . . . Gold Hill, State of Nevada," pp. 72, 77; Virginia City, *Daily Trespass*, Feb. 12, 1867.

35. "Constitution . . . Gold Hill, State of Nevada," pp. 80–90.

36. *Ibid.*

37. *Ibid.*, pp. 94–105.

38. San Francisco, *Mining and Scientific Press*, Oct. 3, 1891; Gold Hill, *Daily News*, Dec. 9, 1867.

39. *Constitution, By-Laws, Order of Business and Rules of Order of the Miners' Union of Virginia, Nevada* (Virginia: Brown and Mahanny, 1879), p. 3.

40. "Constitution . . . Gold Hill, State of Nevada," pp. 121–122; Virginia City, *Daily Trespass*, Aug. 5, 1867; Gold Hill, *Daily News*, Aug. 5, 1867.

41. "Constitution . . . Gold Hill, State of Nevada," pp. 11, 149, 152, 168.

42. *Ibid.*, pp. 13–14, 179–180; *Constitution, By-Laws, Order of Business and Rules of Order of the Miners' Union of Gold Hill, Nev.* (Virginia, Nev.: D. L. Brown, 1885), pp. 15–16; Virginia City, *Chronicle*, July 5, 1917.

43. Angel, *op. cit.*, p. 602; Gold Hill, *Daily News*, Aug. 11, Sep. 22, 1877.

44. "Constitution . . . Gold Hill, State of Nevada," pp. 213–214; Angel, *op. cit.*, p. 261.

45. *Ibid.*, p. 261; Virginia City, *Territorial Enterprise*, Feb. 3, 16, 1878; San Francisco, *Mining and Scientific Press*, Oct. 3, 1891; Lord, *op. cit.*, p. 377; *Constitution . . . of Virginia, Nevada.*, pp. 6, 11.

46. Virginia City, *Territorial Enterprise*, Sept. 9, 1877.

47. This incident and the role of the miner's unions in the anti-Chinese agitation is treated in detail in chapter 5.

48. Gold Hill, *Daily News*, Jan. 16, 22, 25, March 4, 1869; Virginia City, *Territorial Enterprise*, March 5, 1869.

49. Gold Hill, *Daily News*, Feb. 3, 5, 1869.

50. Angel, *op. cit.*, p. 606; *Journals of the Senate and Assembly, Fifth Session of the Nevada State Legislature* (Carson City, 1870); Virginia City, *Territorial Enterprise*, March 6, 11, 1881.

51. Gold Hill, *Daily News*, June 4, July 5, 1867; Virginia City, *Territorial Enterprise*, Sept. 30, 1869; Angel, *op. cit.*, p. 607.

52. Gold Hill, *Daily News*, Jan. 18, 1878; Virginia City, *Chronicle*, quoted in Butte, Montana, *Miner*, Jan. 29, 1878.

53. *Ibid.*

54. Gold Hill, *Daily News*, Oct. 5, 8, 1874; Jan. 15, 1878; Virginia City, *Territorial Enterprise*, Jan. 15, 16, 1878.

55. Gold Hill, *Daily News*, Jan. 17, 1878; Virginia City, *Territorial Enterprise*, Jan. 16, 1878; Virginia City, *Chronicle*, quoted in Butte, Montana, *Miner*, Jan. 29, 1878.

56. Gold Hill, *Daily News*, Jan. 17, 18, 1878; Virginia City, *Territorial Enterprise*, Jan. 18, 1878.

57. *Ibid.*

58. Virginia City, *Territorial Enterprise*, Jan. 18, 1878; Virginia City, *Chronicle*, quoted in Butte, Montana, *Miner*, Jan. 29, 1878.

59. San Francisco, *Mining and Scientific Press*, Aug. 4, 1877; Virginia City, *Territorial Enterprise*, Jan. 20, 26, 1878; New York, *Mining Record*, July 23, 1881; Hailey, Idaho, *Wood River Times*, Jan. 22, 1885; "Agreement entered into between the mining superintendents and the mechanics union, both of Storey County, State of Nevada, July 17th, A.D. 1878, defining mechanical labor, its hours and wages." Two-page manuscript in Special Collections of the University of Nevada Library, Reno.

60. Gold Hill, *Daily News*, Aug. 29, Oct. 26, 1869; Angel, *op. cit.*, pp. 507–508.

61. Gold Hill, *Daily News*, July 14, 1877; Virginia City, *Territorial Enterprise*, Aug. 9, 1877.

62. Virginia City, *Territorial Enterprise*, Aug. 11, 1877; San Francisco, *Stock Exchange*, quoted in the Eureka, Nev., *Daily Sentinel*, Aug. 18, 1877.

63. Virginia City, *Territorial Enterprise*, Aug. 9, 1877; Virginia City, *Chronicle*, Aug. 27, 1877, quoted in San Francisco, *Mining and Scientific Press*, Sept. 1, 1877; Eureka, Nev., *Daily Sentinel*, Aug. 30, 1877.

64. Gold Hill, *Daily News*, Sept. 10, 1877.

65. Carson City, *Daily Appeal*, quoted in San Francisco, *Mining and Scientific Press*, Dec. 11, 1880.

66. Virginia City, *Territorial Enterprise*, Feb. 15, 1881.

67. *Ibid.*, April 2, 1872, March 19, 1881; Gold Hill, *Daily News*, April 1, 3, 1872.

68. Virginia City, *Chronicle*, July 5, 1917; Lord, *op. cit.*, p. 358.

69. *Ibid.*

70. Lord, *op. cit.*, pp. 266, 359; New York, *Engineering and Mining Journal*, Oct. 18, 25, 1890.

THE TURBULENT YEARS

1. Austin, Nev., *Reese River Reveille*, March 9, 1868.
2. *Ibid.*, March 16, 1868; Virginia City, *Daily Trespass*, March 12, 1868; "Constitution and By-Laws, Pledge and Minutes of the Miners' Union of the Town of Gold Hill, State of Nevada," pp. 201, 216. Manuscript in Special Collections of the University of Nevada Library, Reno.
3. The history of the White Pine mines is given in W. Turrentine Jackson, *Treasure Hill: Portrait of a Silver Mining Camp* (Tucson: University of Arizona Press, 1963). The miners' strike is dicussed on pp. 128–138.
4. Gold Hill, *Daily News*, Jan. 8, 1869.
5. *Ibid.*, Jan. 25, 1869.
6. Treasure City, Nev., *White Pine News*, April 20, 21, 22, 23, May 22, 1869.
7. *Ibid.*, April 22, 1869.
8. *Ibid.*, May 15, 22, June 4, July 1, 12, 1869.
9. *Ibid.*, July 12, 1869.
10. Hamilton, Nev., *Inland Empire*, July 14, 1869; Shermantown, Nev., *White Pine Telegram*, July 14, 17, 1869; Treasure City, Nev., *White Pine News*, July 12, 1869.
11. Treasure City, Nev., *White Pine News*, July 12, 13, 14, 1869.
12. Hamilton, Nev., *Inland Empire*, July 28, 1869.
13. Treasure City, Nev., *White Pine News*, July 17, 1869; Hamilton, Nev., *Inland Empire*, July 27, 1869.
14. Treasure City, Nev., *White Pine News*, July 28, 1869; Hamilton Nev., *Inland Empire*, July 28, 1869.
15. Treasure City, Nev., *White Pine News*, July 30, 31, 1869; Hamilton, Nev., *Inland Empire*, July 29, 30, 1869.
16. Treasure City, Nev., *White Pine News*, Aug. 2, 3, 4, 5, 1869; Hamilton, Nev., *Inland Empire*, Aug. 3, 4, 5, 1869.
17. Treasure City, Nev., *White Pine News*, Aug. 5, 6, 1869.
18. *Ibid.*, Aug. 7, 1869.
19. *Ibid.*, Aug. 9, 1869.
20. Treasure City, Nev., *White Pine News*, Aug. 9, 10, 1869; Hamilton, Nev., *Inland Empire*, Aug. 8, 14, 1869.
21. Treasure City, Nev., *White Pine News*, Aug. 9, 1869.
22. James J. Ayers, *Gold and Sunshine: Reminiscences of Early California* (Boston: Richard G. Badger, 1922), pp. 239–240.
23. Austin, Nev., *Reese River Reveille*, June 6, 8, 1872.
24. *Ibid.*, Feb. 19, 1872.
25. *Ibid.*, Feb. 19, 20, 23, 24, 26, March 5, June 26, 1872.
26. *Biennial Report of the State Mineralogist of the State of Nevada for the Years 1871 and 1872* (Carson City, 1873), p. 84; Pioche, Nev., *Record*, Jan. 3, 1873.
27. Pioche, Nev., *Record*, Dec. 27, 1872.
28. *Ibid.*, Dec. 29, 31, 1872, Jan. 1, 4, 1873.
29. *Ibid.*, Jan. 4, 5, 7, March 14, June 4, 1873.
30. Eureka, Nev., *Daily Sentinel*, Sept. 11, 1877; Austin, Nev., *Reese River Reveille*, June 1, 1875.
31. Silver City, Idaho, *Owyhee Avalanche*, Oct. 5, 12, 1867, March 23, 1872.
32. *Ibid.*, March 23, 30, 1872.
33. Los Angeles, *Mining Review*, March 16, 1907.
34. Rossiter W. Raymond, *Statistics of Mines and Mining . . . for the Year 1869* (Washington: G.P.O., 1870), p. 54.

35. Los Angeles, *Mining Review*, March 16, 1907.
36. Raymond, *op. cit.*, pp. 54–55.
37. *Ibid.*, p. 56; Grass Valley, Calif., *Daily National*, April 28, 1869; Jewett V. Reed and A. K. Harcourt, *The Essentials of Occupational Diseases* (Springfield, Ill.: Charles C. Thomas, 1941), p. 76; San Francisco, *Alta California*, May 22, 1869; Virginia City, *Territorial Enterprise*, June 3, 1869.
38. Grass Valley, Calif., *Daily National*, April 22, 26, May 5, 10, 11, 12, 1869; San Francisco, *Mining and Scientific Press*, May 1, 8, 26, 1869.
39. Grass Valley, Calif., *Daily National*, April 28, May 10, 1869.
40. *Ibid.*, May 10, 11, 1869.
41. *Ibid.*, May 17, 1869.
42. *Ibid.*, May 14, 17, 18, 26, 1869; *Constitution, By-Laws, Order of Business and Rules of Order of the Miners' Union of Grass Valley, Cal., Organized May 1869* (Grass Valley: Daily National, 1869).
43. *Ibid.*, May 22, 1869.
44. *Ibid.*, May 25, 1869; Nevada City, Calif., *Gazette*, May 25, 1869; Virginia City, *Territorial Enterprise*, May 27, 1869.
45. Grass Valley, Calif., *Daily National*, May 26, 1869.
46. *Ibid.*, May 27, 1869; Nevada City, Calif., *Gazette*, May 25, 1869.
47. Grass Valley, Calif., *Daily Naitonal*, May 28, 1869.
48. *Ibid.*, July 12, 14, 1869.
49. *Ibid.*, July 15, 21, 1869; San Francisco, *Alta California*, July 19, 1869.
50. Grass Valley, Calif., *Daily National*, July 13, 20, Aug. 7, 1869; Raymond, *op. cit.*, p. 54.
51. Grass Valley, Calif., *Daily National*, Aug. 26, Sept. 6, 28, Oct. 26, Nov. 13, 1869.
52. Gold Hill, *Daily News*, March 20, 1872.
53. *Ibid.*, Feb. 10, 1872; Grass Valley, Calif., *Republican*, Feb. 20, 1872.
54. Grass Valley, Calif., *Republican*, Feb. 20, 21, 23, 29, 1872.
55. Grass Valley, Calif., *Republican*, March 1, 6, 7, 1872; Gold Hill, *Daily News*, March 4, 1872.
56. Grass Valley, Calif., *Republican*, March 5, 8, 1872.
57. *Ibid.*, March 10, 17, 1872.
58. Rossiter W. Raymond, *Statistics of Mines and Mining . . . for the year 1871* (Washington: G.P.O., 1872), pp. 83–87; *Annual Report of the Amador Mining Company* (San Francisco: Stock Report, 1871); Jackson, Calif., *Amador Dispatch*, Sept. 3, 1870; San Francisco, *Chronicle*, July 18, 1871.
59. San Francisco, *Bulletin*, Aug. 2, 1871.
60. *Ibid.*; Jackson, Calif., *Amador Dispatch*, Aug. 6, 13, 1870, Feb. 18, 25, 1871; Jackson, Calif., *Amador Ledger*, Aug. 20, 1870; Sacramento, *Daily Union*, June 7, 1871; Jesse D. Mason, *History of Amador County* (Oakland: Thompson and West, 1881), pp. 112, 287.
61. Jackson, Calif., *Amador Ledger*, Aug. 20, 1870; Jackson, Calif., *Amador Dispatch*, Sept. 3, 10, 1870.
62. Mason, *op. cit.*, p. 149; San Francisco, *Bulletin*, Aug. 2, 1869.
63. Sacramento, *Daily Union*, June 7, 1871.
64. *Ibid.*, Mason, *op. cit.*, p. 112.
65. Sacramento, *Daily Union*, June 7, 1871.
66. Jackson, Calif., *Amador Dispatch*, June 3, 1871; Jackson, Calif., *Amador Ledger*, June 3, 1871; San Francisco, *Chronicle*, June 22, 1871; San Francisco, *Alta California*, June 2, 1871, Oct. 10, 1878.
67. Jackson, Calif., *Amador Dispatch*, June 3, 10, 1871; Jackson, Calif., *Amador Ledger*, June 3, 1871.
68. Jackson, Calif., *Amador Dispatch*, June 10, 24, 1871; San Francisco, *Alta California*, June 2, 20, 1871.

69. *Report of the Adjutant-General of the State of California for the Years 1870 and 1871* (Sacramento: State Printer, 1871), p. 27.

70. *Ibid.*, "The California Recollections of Caspar T. Hopkins," *California Historical Society Quarterly*, 26 (1947), 353.

71. *Report of the Adjutant-General ... 1870 and 1871*, p. 30; San Francisco, *Alta California*, June 22, 23, 1871; San Francisco, *Chronicle*, June 22, 23, 1871; San Francisco, *Bulletin*, June 22, 23, 1871.

72. San Francisco, *Chronicle*, June 22, 23, 1871; San Francisco, *Alta California*, June 23, 1871.

73. San Francisco, *Bulletin*, June 23, 1871; San Francisco, *Chronicle*, June 24, 1871.

74. Sacramento, *Reporter*, June 24, 1871; *Report of the Adjutant-General ... 1870 and 1871*, p. 117; San Francisco, *Chronicle*, June 25, 1871; San Francisco, *Alta California*, June 25, 27, 1871.

75. San Francisco, *Alta California*, June 25, 27, 1871.

76. *Ibid.*, June 25, 26, 27, 1871; San Francisco, *Chronicle*, June 25, 27, 1871; San Francisco, *Bulletin*, June 26, 1871.

77. San Francisco, *Alta California*, June 28, 1871.

78. San Francisco, *Chronicle*, June 27, 1871; San Francisco, *Bulletin*, June 27, 1871.

79. San Francisco, *Chronicle*, June 27, 28, 1871; San Francisco, *Alta California*, June 28, 1871.

80. Sacramento, *Reporter*, June 30, 1871; San Francisco, *Chronicle*, June 28, 1871; San Francisco, *Alta California*, June 29, 30, July 4, 1871.

81. San Francisco, *Alta California*, June 28, 29, July 1, 1871; San Francisco, *Chronicle*, June 29, 30, 1871; San Francisco, *Bulletin*, June 28, 1871.

82. San Francisco, *Chronicle*, July 6, 11, 1871.

83. Jackson, Calif., *Amador Dispatch*, July 15, 1871.

84. *Ibid.*; *Report of the Adjutant-General ... 1870 and 1871*, p. 31.

85. San Francisco, *Chronicle*, July 19, 25, 1871; San Francisco, *Alta California*, July 19, 1871; *Annual Report of the Amador Mining Company* (San Francisco: Stock Report, 1872), pp. 15–18; *Report of the Adjutant-General ... 1870 and 1871*, p. 32.

86. San Francisco, *Chronicle*, July 27, 1871; San Francisco, *Alta California*, July 25, 27, 28, 1871; San Francisco, *Bulletin*, July 26, 27, 1871; Jackson, Calif., *Amador Dispatch*, Aug. 12, 1871.

87. San Francisco, *Chronicle*, July 29, Aug. 1, 1871; San Francisco, *Alta California*, July 29, 31, 1871.

88. San Francisco, *Alta California*, Aug. 11, 1871.

89. Antioch, Calif., *Ledger*, Sept. 2, 1871.

90. Jackson, Calif., *Amador Ledger*, Sept. 16, 1871.

91. Central City, Colo., *Daily Register*, Nov. 11, 15, 1873; Denver, *Rocky Mountain News*, Nov. 12, 1873; Caroline Bancroft, *Gulch of Gold: A History of Central City, Colorado* (Denver: Sage Books, 1958), pp. 252–255.

92. Central City, Colo., *Daily Register*, Nov. 13, 14, 15, 17, 1873; Denver, *Rocky Mountain News*, Nov. 12, 13, 15, 16, 18, 1873.

93. Central City, Colo., *Daily Register*, Nov. 17, Dec. 3, 1873; Denver, *Rocky Mountain News*, Nov. 12, 18, 22, 1873; Bancroft, *op. cit.*, p. 255.

UNION AGAINST THE CHINESE

1. Various aspects of the history of the Chinese in the western states have been treated in Alexander P. Saxton, *The Indispensable Enemy: Labor and the*

Anti-Chinese Movement in California (Berkeley and Los Angeles: University of California Press, 1971); Ping Chiu, *Chinese Labor in California, 1850–1880: An Economic Study* (Madison: State Historical Society of Wisconsin, 1963); Elmer C. Sandmeyer, *The Anti-Chinese Movement in California* (Urbana: University of Illinois Press, 1939); Gunther Barth, *Bitter Strength: A History of the Chinese in the United States, 1850–1870* (Cambridge, Mass.: Harvard University Press, 1964); Gary P. Be Dunnah, "A History of the Chinese in Nevada: 1853–1904," Unpublished M.A. Thesis, University of Nevada, Reno, 1966; Rose Hum Lee, "The Growth and Decline of Chinese Communities in the Rocky Mountain Region," Unpublished Ph.D. Dissertation, University of Chicago, 1947; Fern Coble Trull, "The History of the Chinese in Idaho from 1864 to 1910," Unpublished M.A. Thesis, University of Oregon, 1946; and Gerald E. Rudolph, "The Chinese in Colorado, 1869–1911," Unpublished M.A. Thesis, University of Denver, 1964.

2. Rossiter W. Raymond, *Statistics of Mines and Mining . . . for the Year 1870* (Washington: G.P.O. 1872), pp. 3–6; Gold Hill, *Daily News*, July 6, 1867.

3. Grass Valley, Calif., *Daily National*, May 25, 31, 1869.

4. Raymond, *op. cit.*, p. 6.

5. Unionville, Nev., *Humboldt Register*, Jan. 23, 1869; "Constitution and By-Laws, Pledge and Minutes of the Miners' Union of the Town of Gold Hill, State of Nevada," p. 18, Manuscript in Special Collections of the University of Nevada Library, Reno; Myron Angel, ed., *History of Nevada* (Oakland: Thompson and West, 1881), p. 447.

6. Unionville, Nev., *Humboldt Register*, Jan. 9, 23, 1869.

7. *Ibid.*, Jan. 16, 23, 1869.

8. *Ibid.*, Jan. 16, 23, Feb. 13, 1869 (quotes *Carson Appeal*); Virginia City, *Territorial Enterprise*, March 17, 1869.

9. *Treaties, Conventions, International Acts, Protocols and Agreements between the United States of America and Other Powers, 1776–1909*, compiled by William M. Malloy (Washington: G.P.O., 1910), I, 236; Unionville, Nev., *Humboldt Register and Workingman's Advocate*, April 17, 1869.

10. *Ibid.*, April 24, May 1, 22, 1869.

11. Gold Hill, *Daily News*, Jan. 16, 22, 25, March 4, 1869; Virginia City, *Territorial Enterprise*, Feb. 26, March 5, 1869.

12. Unionville, Nev., *Humboldt Register and Workingman's Advocate*, Feb. 13, 1869; John Koontz, *Political History of Nevada* (5th ed.; Carson City, 1965), pp. 119, 123.

13. Unionville, Nev., *Humboldt Register and Workingman's Advocate*, Feb. 13, April 10, May 1, 1869.

14. *Ibid.*, Feb. 13, 20, March 6, 1869.

15. *Ibid.*, May 15, 1869.

16. Virginia City, *Territorial Enterprise*, May 29, June 2, 1869.

17. *Ibid.*, July 7, 1869.

18. *Ibid.*

19. *Ibid.*

20. *Ibid.*, Aug. 3, 1869.

21. *Ibid.*

22. Unionville, Nev., *Humboldt Register and Workingman's Advocate*, May 22, 1869; Treasure City, Nev., *White Pine News*, July 17, 1869; San Francisco, *Mining and Scientific Press*, July 24, Aug. 14, 1869.

23. Eliot Lord, *Comstock Mining and Miners* (Washington: G.P.O., 1883), p. 355.

24. Gold Hill, *Daily News*, Sept. 29, 30, 1869; Virginia City, *Territorial Enterprise*, Sept. 30, 1869.

25. Gold Hill, *Daily News*, Sept. 30, Oct. 2, 4, 7, 1869; Virginia City, *Territorial Enterprise*, Sept. 30, 1869.

26. Virginia City, *Territorial Enterprise*, Sept. 30, 1869.

27. Gold Hill, *Daily News*, Oct. 1, 5, 1869.

28. *Ibid.*, Oct. 7, 1869.

29. *Ibid.*

30. Grass Valley, Calif., *Daily National*, May 27, Sept. 6, 28, Oct. 18, 19, 27, 1869; Nevada City, Calif., *Gazette*, May 25, 1869; Jackson, Calif., *Amador Ledger*, Aug. 20, 1870; Jackson, Calif., *Amador Dispatch*, July 15, 1871; San Francisco, *Mining and Scientific Press*, April 19, 26, 1873.

31. Silver City, Idaho, *Owyhee Avalanche*, March 1, 8, 1873.

32. *Ibid.*, March 8, 15, May 3, 1873.

33. Boulder, Colo., *Boulder County News*, March 20, April 10, 17, 1874 and Denver, Colo., *Rocky Mountain News*, April 2, 1874, cited in Duane A. Smith, "The Caribou—A Forgotten Mine," *Colorado Magazine*, 39 (1962), 52.

34. North San Juan, Calif., *Times*, March 11, 1876.

35. *Ibid.*; San Francisco, *Mining and Scientific Press*, March 26, 1876.

36. Grass Valley, Calif., *Foothill Tidings*, March 17, 1877; North San Juan, Calif., *Times*, Dec. 9, 1876, March 31, 1877.

37. Nevada City, Calif., *Daily Transcript*, Jan. 27, 30, Feb. 21, March 28, 1877.

38. North San Juan, Calif., *Times*, Sept. 22, 29, 1877, June 5, 1880; Harry L. Wells, ed., *History of Nevada County, California* (Oakland: Thompson and West, 1880), p. 62.

39. Nevada City, Calif., *Daily Transcript*, March 14, 20, 1877; San Francisco, *Mining and Scientific Press*, June 3, 1882.

40. Virginia City, *Territorial Enterprise*, May 7, 9, 1882; San Francisco, *Mining and Scientific Press*, April 8, Sept. 30, 1882; Downieville, Calif., *Mountain Messenger*, Nov. 10, 1883; Ruby Hill, Nev., *Mining News*, Nov. 26, 1883.

41. Deadwood, S.D., *Black Hills Daily Times*, Feb. 19, 1878; San Bernardino, Calif., *Times*, May 13, 1876.

42. Belmont, Nev., *Courier*, May 5, 1876; Eureka, Nev., *Daily Sentinel*, March 9, April 30, May 6, 11, 1876.

43. Eureka, Nev., *Daily Sentinel*, June 10, 1876, Jan. 1, 1877.

44. Bodie, Calif., *Daily Free Press*, May 25, 26, 1881.

45. *Ibid.*, May 27, 28, 1881.

46. *Ibid.*, May 27, 28, June 6, 1881 (quoting San Francisco *Stock Report*).

47. Butte, Montana, *Weekly Miner*, Dec. 27, 1881.

48. Ira B. Cross, *A History of the Labor Movement in California* (Berkeley: University of California Press, 1935), pp. 88 ff.

49. Sandmeyer, *op. cit.*, pp. 67–74.

50. *Ibid.*; *Treaties . . .* , *op. cit.*, I, 237–239.

51. San Francisco, *Alta California*, April 10, 25, 1882.

52. *Ibid.*, April 25, 26, 27, 28, 29, 30, May 5, 1882.

53. Cross, *op. cit.*, pp. 139 ff.; Virginia City, *Territorial Enterprise*, May 9, 11, 15, 18, 23, 1882.

54. Isaac H. Bromley, *The Chinese Massacre at Rock Springs, Wyoming Territory, September 2, 1885* (Boston: Franklin Press, 1886); Paul Crane and Alfred Larson, "The Chinese Massacre," *Annals of Wyoming*, 12 (Jan. 1940), 47–55.

THE BONANZA YEARS

1. Eureka, Nev., *Daily Sentinel*, Sept. 23, 1877; *Constitution, By-Laws, Order of Business and Rules of Order of the Miners' Union of Gold Hill, Nev.* (Virginia, Nev.: D. L. Brown, 1885), p. 6.

2. Eureka, Nev., *Daily Sentinel*, March 14, 15, 31, April 6, 15, May 2, 1876, Nov. 11, 1877.

3. *Ibid.*, July 29, 1877; Virginia City, *Territorial Enterprise*, Aug. 7, 1877.

4. Austin, Nev., *Reese River Reveille*, Nov. 9, 13, 1877; Eureka, Nev., *Daily Sentinel*, Sept. 13, 23, Nov. 2, 11, 1877.

5. Virginia City, *Territorial Enterprise*, quoted in Butte, Mont., *Miner*, Jan. 15, 1878; San Francisco, *Mining and Scientific Press*, Dec. 25, 1880.

6. Bodie, Calif., *Weekly Standard*, Jan. 2, 23, June 19, 1878; Bodie, *Morning News*, July 10, 1879; Bodie, *Daily Free Press*, April 30, May 19, 23, 1880; Lundy, Calif., *Homer Mining Index*, Feb. 5, 26, 1881; Aurora, Nev., *Esmeralda Herald*, March 18, 25, April 8, 1882.

7. San Francisco, *Mining and Scientific Press*, Dec. 22, 1877, Aug. 6, 1881, June 3, Sept. 30, 1882, Nov. 3, 17, 1883; Downieville, Calif., *Mountain Messenger*, July 9, 1881; Nevada City, Calif., *Daily Transcript*, March 14, 26, 1877, June 3, 17, 1879.

8. Deadwood, S. Dak., *Black Hills Weekly Pioneer*, Nov. 3, 1877, Jan. 1, 1882; Deadwood, *Black Hills Daily Times*, Dec. 22, 1877, Jan. 15, Feb. 19, Oct. 3, 22, 1878.

9. Butte, Mont., *Miner*, June 11, 18, July 2, 9, 23, 30, Aug. 6, Oct. 8, 1878; Butte, *Daily Inter Mountain*, June 13, 1891; K. Ross Toole, "A History of the Anaconda Copper Mining Company: A Study in the Relationship between a State and Its People and a Corporation, 1880–1950." Unpublished Ph.D. Dissertation, University of California, Los Angeles, 1954, pp. 156–158. The early history of the Butte union is discussed in detail in chapter 7.

10. See the detailed discussion of the Leadville miners' unions in the latter part of this chapter.

11. Silver Reef, Utah, *Miner*, Feb. 14, Aug. 21, 1880; Park City, Utah, *Mining Record*, June 5, 1880, July 16, 1881; Hailey, Idaho, *Wood River Times*, Aug. 17, Sept. 7, 1881, March 8, July 12, 27, 1882, March 14, 17, 1885; New York, *Engineering and Mining Journal*, Nov. 17, 24, 1883; Myron Angel, ed., *History of Nevada* (Oakland, Calif.: Thompson and West, 1881), p. 657; Tuscarora, Nev., *Times-Review*, June 3, 5, 9, July 5, 1880; Eureka, Nev., *Daily Sentinel*, July 1, 1880; Austin, Nev., *Reese River Reveille*, Feb. 8, 1881; Tucson, Ariz., *Daily Star*, April 30, May 17, July 6, 1884; *Constitution and By-Laws of the Globe City Union of Globe, Arizona, Organized May 14th, 1884* (Globe: Arizona Silver Belt Print, 1884).

12. Austin, Nev., *Reese River Reveille*, May 29, June 1, 1875.

13. Eureka, Nev., *Daily Sentinel*, March 2, 28, April 6, 1876.

14. *Ibid.*, April 26, June 25, 29, 1876.

15. Virginia City, *Daily Chronicle*, April 27, May 2, 1877; Virginia City, *Territorial Enterprise*, May 3, 1877; Eureka, Nev., *Daily Sentinel*, May 3, 1877.

16. Silver City, *Idaho Avalanche*, July 8, 1876.

17. *Ibid.*, July 8, 22, Sept. 16, 1876.

18. San Francisco, *Mining and Scientific Press*, May 29, 1880; Gold Hill, *Daily News*, Nov. 22, 1877; New York, *Times*, Nov. 11, 13, 1877; Estelline Bennett, *Old Deadwood Days* (New York: J. H. Sears and Co., 1928), pp. 51–53.

19. Eureka, Nev., *Daily Sentinel*, Nov. 4, 1879; Lundy, Calif., *Homer Mining Index*, June 26, 1880; San Francisco, *Mining and Scientific Press*, Aug. 27, 1881; New York, *Mining Record*, June 12, 1880; Bodie, Calif., *Weekly Standard-News*, Aug. 24, 1881.

20. Bodie, Calif., *Morning News*, Aug. 24, 26, 27, 1879; Virginia City, *Territorial Enterprise*, Aug. 24, 26, 27, 28, 29, Sept. 17, 1879.

21. Bodie, Calif., *Morning News*, Sept. 21, 23, 1879; Aurora, Nev., *Esmeralda Herald*, Sept. 27, Oct. 11, 1879; Virginia City, *Territorial Enterprise*, Sept. 23, 26, 1879.

22. Bodie, Calif., *Daily Free Press*, Sept. 22, 1879, quoted in Aurora, Nev., *Esmeralda Herald*, Sept. 27, 1879; Virginia City, *Territorial Enterprise*, Sept. 26, 1879.

23. Eureka, Nev., *Daily Sentinel*, Oct. 10, Nov. 4, 1879.

24. *Ibid.*, Oct. 17, 1879; Bodie, Calif., *Morning News*, Oct. 4, 1879; Aurora, Nev., *Esmeralda Herald*, Oct. 11, 1879.

25. Bodie, Calif., *Morning News*, Oct. 2, 1879; Bodie, *Weekly Standard-News*, Aug. 24, 31, 1881.

26. Leadville, Colo., *Daily Democrat*, June 9, 1880; Nevada City, Calif., *Daily Transcript*, May 19, 1888; Eureka, Nev., *Daily Sentinel*, Feb. 16, 17, 18, 19, 20, 22, 25, 1881; Eureka, *Daily Leader*, Feb. 16, 18, 19, 21, 1881.

27. Eureka, Nev., *Daily Leader*, Feb. 19, 1881.

28. Bodie, Calif., *Weekly Standard*, Feb. 15, 22, 1879.

29. Eureka, Nev., *Daily Sentinel*, March 29, 1881; Austin, Nev., *Reese River Reveille*, April 26, 1881; Phillip I. Earl, "Nevada's Italian War, July–September 1879," *Nevada Historical Society Quarterly*, 12 (1969), 51–87.

30. Independence, Calif., *Inyo Independent*, Feb. 23, May 25, June 1, 1878; Bodie, Calif., *Weekly Standard*, May 29, 1878.

31. Leadville, Colo., *Daily Democrat*, March 9, June 1, 18, 1880; Leadville, *Weekly Herald*, April 3, 1880; *Tenth Census of the United States* (Washington: G.P.O., 1885), XIV, 486–491, Mining Laws.

32. The Leadville strike has been discussed in more or less detail by many writers: R. G. Dill, "History of Lake County" in *History of the Arkansas Valley, Colorado* (Chicago: O. L. Baskin and Co., 1881), pp. 239–245; R. G. Dill, *The Political Campaigns of Colorado* (Denver: Arapahoe Publishing Co., 1895), pp. 50–55; Carlyle C. Davis, *Olden Times in Colorado* (Los Angeles: Phillips Publishing Co., 1916), pp. 248–261; Joseph R. Buchanan, *The Story of a Labor Agitator* (New York: Outlook Co., 1903), pp. 8–36; *A Report of Labor Disturbances in the State of Colorado from 1880 to 1904, Inclusive* (*58th Cong., 3rd. Sess., Sen. Doc. No. 122*) (Washington: G.P.O., 1905), pp. 69–74; Vernon H. Jensen, *Heritage of Conflict* (Ithaca: Cornell University Press, 1950), pp. 19–24; Don Griswold and Jean Griswold, *The Carbonate Camp Called Leadville* (Denver: University of Colorado Press, 1951), pp. 180–199; Charles M. Hough, "Leadville, Colorado, 1878 to 1898: A Study in Unionism," Unpublished M.A. Thesis, University of Colorado, 1958, pp. 44–60; Paul T. Bechtol, Jr., "The 1880 Labor Dispute in Leadville," *Colorado Magazine*, 47 (1970), 312–325; and Duane A. Smith, *Horace Tabor: His Life and the Legend* (Boulder: Colorado Associated University Press, 1973). Only Dill (1895) and Smith have appreciated the important role of the Chrysolite stock promotion scheme in the initiation and perpetuation of the strike, and the author is indebted to Professor Smith for calling his attention to this role.

33. Leadville, Colo., *Daily Democrat*, March 16, April 1, 7, 13, 16, May 5, 1880; Asbury Harpending, *The Great Diamond Hoax and Other Stirring Incidents in the Life of Asbury Harpending* (San Francisco: James H. Barry Co., 1913, and Norman: University of Oklahoma Press, 1958); Bruce A. Woodard, *Diamonds in the Salt* (Boulder: Pruett Press, 1967); one has to read between the lines in the last two items.

34. New York, *Engineering and Mining Journal*, May 15, 1880, Sept. 4, 1920.

35. *Ibid.*, June 5, 1880, June 1941 p. 51; New York, *Times*, April 2, 1881; San Francisco, *Mining and Scientific Press*, Oct. 11, 1919.

36. Leadville, Colo., *Daily Chronicle*, May 26, 1880, quoted in Eureka, Nev., *Daily Sentinel*, June 1, 1880.

37. *Ibid.*; Leadville, *Daily Democrat*, May 27, 1880; Leadville, *Weekly Herald*, May 29, 1880.

38. Leadville, Colo., *Weekly Herald*, May 29, June 5, 1880; Leadville, *Daily Democrat*, March 28, May 27, 1880; Davis, *op. cit.*, p. 249.

39. Leadville, Colo., *Weekly Herald*, May 29, 1880.

40. *Ibid.*

41. *Ibid.*, June 5, 1880.

42. Leadville, Colo., *Daily Democrat*, May 29, 30, 1880.

43. *Ibid.*, May 29, 1880.

44. *Ibid.*

45. Leadville, *Weekly Herald*, May 29, June 5, 1880; Leadville, *Daily Democrat*, May 29, June 3, 5, 15, 17, 1880.

46. *Ibid.*, June 1, 1880; Leadville, *Weekly Herald*, June 5, 1880; New York, *Engineering and Mining Journal*, June 5, 1880.

47. Leadville, *Weekly Herald*, June 5, 1880.

48. *Ibid.*; Leadville, *Daily Democrat*, June 1, 1880.

49. Leadville, *Daily Democrat*, June 1, 2, 5, 1880.

50. *Ibid.*, June 3, 5, 1880.

51. *Ibid.*, June 8, 9, 1880.

52. *Ibid.*, June 11, 12, 1880.

53. *Ibid.*; Davis, *op. cit.*, pp. 248–249.

54. New York, *Engineering and Mining Journal*, June 5, 12, 19, 26, 1880; Leadville, *Weekly Herald*, June 12, 1880; Leadville, *Daily Democrat*, June 11, 1880.

55. Davis, *op. cit.*, pp. 250–255.

56. Leadville, *Daily Democrat*, June 12, 1880.

57. *Ibid.*

58. *Ibid.*, June 13, 1880; Dill, "History of Lake County," p. 244; Buchanan, *op. cit.*, pp. 21 ff.

59. *Biennial Report of the Adjutant General of the State of Colorado for the Term Ending December 31st 1880* (Denver: Tribune Publishing Co., 1881), p. 34; Leadville, *Daily Democrat*, June 13, 1880; New York, *Tribune*, June 14, 1880.

60. Leadville, *Weekly Herald*, June 19, 1880; *First Biennial Report of Bureau of Labor Statistics of State of Colorado* (Denver, 1888), p. 138.

61. *Ibid.*

62. Leadville, *Daily Democrat*, June 15, 1880; Leadville, *Weekly Herald*, June 19, 1880; Buchanan, *op. cit.*, p. 26.

63. Leadville, *Daily Democrat*, June 16, 17, 18, 1880.

64. *Ibid.*, June 17, 18, 24, 25, 1880; Leadville, *Weekly Herald*, June 19, 1880; Thomas B. Corbett and John H. Ballanger, *Second Annual Leadville City Directory* (Leadville, 1881), p. 219.

65. Leadville, *Daily Democrat*, June 18, 19, 22, 23, 24, 1880; *A Report of Labor Disturbances . . .* p. 360.

66. Leadville, *Weekly Herald*, July 31, Aug. 21, 1880; New York, *Engineering and Mining Journal*, July 24, Aug. 7, Sept. 25, 1880.

67. New York, *Engineering and Mining Journal*, July 31, 1880; Leadville, *Weekly Herald*, Aug. 21, 1880; Bodie, Calif., *Weekly Standard-News*, Aug. 24, 1881.

68. New York, *Engineering and Mining Journal*, Aug. 7, 1880; New York, *Times*, April 2, 1881.

69. New York, *Engineering and Mining Journal*, June through August, 1880; San Francisco, *Mining and Scientific Press*, May 14, Aug. 27, 1881; Spokane, Wash., *Spokane Review*, May 17, 18, 1892; Denver, *Miners' Magazine*, Oct. 22, 1903.

THE FOUR DOLLAR FIGHT

1. Virginia City, *Territorial Enterprise*, Feb. 15, March 19, 1881; Virginia City, *Footlight* quoted in Silver City, *Idaho Avalanche*, March 26, 1881; Eureka, Nev., *Daily Sentinel*, Feb. 25, 1881.

2. New York, *Mining Record*, July 2, 1881.

3. Virginia City, *Territorial Enterprise*, May 12, 21, 1882.

4. New York, *Mining Record*, Jan. 29, 1881; Austin, Nev., *Reese River Reveille*, Feb. 4, 26, 1881.

5. Austin, Nev., *Reese River Reveille*, Feb. 4, 8, 1881; New York, *Mining Record*, March 26, 1881.

6. Austin, Nev., *Reese River Reveille*, Feb. 10, March 10, 21, 1881; Bodie, Calif., *Weekly Standard-News*, March 16, 1881; New York, *Mining Record*, March 26, 1881.

7. Eureka, Nev., *Daily Sentinel*, March 13, 1881; Austin, Nev., *Reese River Reveille*, March 11, 12, 1881.

8. Eureka, Nev., *Daily Sentinel*, March 13, 1881; Austin, Nev., *Reese River Reveille*, March 14, 19, April 14, May 21, 1881; Virginia City, *Territorial Enterprise*, April 16, 1881.

9. San Francisco, *Mining and Scientific Press*, Jan. 15, Feb. 12, 1881; Salt Lake City, *Daily Tribune*, Jan. 1, Feb. 4, 8, 9, 1881; Silver Reef, Utah, *Miner*, Feb. 2, April 27, 1881.

10. San Francisco, *Mining and Scientific Press*, March 26, 1881.

11. Salt Lake City, *Daily Tribune*, March 1, 8, 1881.

12. Austin, Nev., *Reese River Reveille*, April 11, 1881; Eureka, Nev., *Daily Sentinel*, March 17, 22, 1881; Mark A. Pendleton, "Memories of Silver Reef," *Utah Historical Quarterly*, 3 (Oct. 1930), 110–111; Silver Reef, *Miner*, Sept. 10, 1881.

13. Austin, Nev., *Reese River Reveille*, March 26, 1881; Eureka, Nev., *Daily Sentinel*, March 16, 30, 1881; San Francisco, *Mining and Scientific Press*, April 2, 1881.

14. Silver Reef, *Miner*, April 27, 1881.

15. *Ibid.*

16. Virginia City, *Territorial Enterprise*, May 12, 21, 1882; New York, *Times*, May 29, 1882; New York, *Engineering and Mining Journal*, May 13, 1882.

17. *Ibid.*, Jan. 14, 1882; Jan. 20, Feb. 3, 1883; Feb. 2, 9, 1884, and weekly bullion market prices 1884–1885.

18. San Francisco, *Exchange*, quoted in Ruby Hill, Nev., *Mining News*, March 26, 1883.

19. San Francisco, *Mining and Scientific Press*, Jan. 12, 1882.

20. *Ibid.*, Dec. 27, 1884.

21. New York, *Engineering and Mining Journal*, Sept. 8, Nov. 10, 17, 24, Dec. 1, 8, 22, 1883; *Third Annual Report of the Commissioner of Labor, 1887, Strikes and Lockouts* (Washington: G.P.O., 1888), pp. 52–55.

22. William P. Blake, *Tombstone and Its Mines* (New York: Cheltenham Press, 1902); Tuscarora, Nev., *Times-Review*, Nov. 22, 1880; Hailey, Idaho, *Wood River Times*, May 3, 1884; Tucson, *Arizona Daily Star*, April 30, 1884; Tucson, *Arizona Weekly Citizen*, May 3, 24, 1884; San Francisco, *Mining and Scientific Press*, May 10, 1884, Jan. 24, 1885.

23. Tucson, *Arizona Daily Star*, May 1, 1884.

24. *Ibid.*, May 4, 9, 11, 1884; Tucson, *Arizona Weekly Citizen*, May 10, 17, 1884; Eureka, Nev., *Daily Sentinel*, May 7, 1884; San Francisco, *Alta California*, May 11, 24, 1842.

25. Tucson, *Arizona Daily Star*, May 13, 1884; Tucson, *Arizona Weekly Citizen*, May 17, 24, 1884.

26. Tucson, *Arizona Daily Star*, May 17, June 17, July 6, 1884.

27. *Ibid.*, June 12, 1884.

28. Tombstone, *Epitaph*, June 21, 1884, quoted in *ibid.*, June 24, 1884.

29. Tucson, *Arizona Daily Star*, Aug. 1, 1884.

30. *Ibid.*, Aug. 2, 1884.

31. *Ibid.*, Aug. 6, 7, 1884; Tucson, *Arizona Weekly Citizen*, Aug. 9, 1884; Tombstone, *Epitaph*, Aug. 7, 1884, quoted in Eureka, Nev., *Daily Sentinel*, Aug. 16, 1884.

32. Tucson, *Arizona Weekly Citizen*, Aug. 9, 16, 1884; Tucson, *Arizona Daily Star*, Aug. 10, 1884.

33. Tombstone, *Epitaph*, Aug. 9, 1884, quoted in Tucson, *Arizona Weekly Citizen*, Aug. 16, 1884, and in Tucson, *Arizona Daily Star*, Aug. 10, 1884.

34. "Report of the Governor of Arizona," in *Report of the Secretary of the Interior* (Washington: G.P.O., 1884) (48th Cong., 2d Sess. Ex. Doc. 1, Pt. 5), pp. 523–524; Tucson, *Arizona Daily Star*, Aug. 14, 1884.

35. Tucson, *Arizona Daily Star*, Aug. 28, 1884.

36. *Ibid.*, Aug. 29, 1884; New York, *Engineering and Mining Journal*, Aug. 30, Oct. 18, 1884, Jan. 17, Feb. 25, 1885; San Francisco, *Mining and Scientific Press*, Dec. 13, 1884, Jan. 17, Feb. 7, 1885.

37. Hailey, Idaho, *Wood River Times*, March 25, 1885; *Great Register of the County of Mono for the Year 1884* (Bridgeport, Cal.: Chronicle-Union Print, 1884), p. 15.

38. Hailey, *Wood River Times*, March 22, 1882, June 21, 1884; San Francisco, *Mining and Scientific Press*, Jan. 31, 1885.

39. Hailey, *Wood River Times*, June 21, July 21, 25, 1884, Jan. 22, 1885.

40. *Ibid.*, Jan. 19, 20, Feb. 6, 7, 14, 16, 1885.

41. *Ibid.*, Feb. 6, 14, 16, 18, 20, 1885.

42. *Ibid.*, Feb. 18, 19, 1885.

43. *Ibid.*, Feb. 12, 23, 1885.

44. *Ibid.*, Feb. 24, 25, 1885.

45. *Ibid.*, Feb. 28, 1885.

46. *Ibid.*, Feb. 28, March 4, 14, 1885.

47. *Ibid.*, March 2, 3, 4, 7, 11, 13, 14, 18, 1885.

48. *Ibid.*, March 16, 17, 1885.

49. *Ibid.*, March 28, 1885.

50. *Ibid.*, March 19, 27, 1885.

51. *Ibid.*, March 25, 26, 27, 28, 30, 1885.

52. *Ibid.*, March 18, 19, 1885.

53. *Ibid.*, March 18, 20, 24, 1885; Boise, Idaho, *Tri-Weekly Statesman*, March 19, 1885.

54. Hailey, *Wood River Times*, March 23, 28, 1885; Boise, Idaho, *Tri-Weekly Statesman*, March 24, 1885.

55. Hailey, *Wood River Times*, March 23, 24, 25, 31, April 2, 1885; Bellevue, Idaho, *Chronicle*, March 28, 1885, quoted in Silver City, *Idaho Avalanche*, April 4, 1885.

56. Hailey, *Wood River Times*, March 24, April 9, 11, 1885.

57. Eureka, Nev., *Daily Sentinel*, May 22, 1885; Hailey, *Wood River Times*, April 8, 23, 25, May 19, 1885.

58. Hailey, *Wood River Times*, May 9, Nov. 2, 1885; Nevada City, Calif., *Daily Transcript*, Feb. 12, 1888; San Francisco, *Mining and Scientific Press*, Feb. 11, 1888.

59. Eureka, Nev., Daily Leader, Aug. 5, 6, 1884; San Francisco, *Mining and*

Scientific Press, Aug. 16, 1884; New York, *Engineering and Mining Journal*, Sept. 13, 1884.

60. Ruby Hill, Nev., *Mining News*, July 28, 1884; Eureka, Nev., *Daily Sentinel*, June 23, 1885.

61. Eureka, Nev., *Daily Leader*, March 23, 1885; Eureka, *Daily Sentinel*, March 24, 1885.

62. Eureka, *Daily Sentinel*, April 5, 19, May 7, 1885; Salt Lake City, *Daily Tribune*, March 29, 1885.

63. Eureka, *Daily Sentinel*, April 7, 1885; Hailey, *Wood River Times*, April 9, 1885.

64. Eureka, *Daily Sentinel*, June 27, 1885, June 22, Oct. 6, 1886.

65. *Ibid.*, May 27, 1885; Hailey, *Wood River Times*, Sept. 25, 1885; San Francisco, *Mining and Scientific Press*, June 27, 1885; New York, *Engineering and Mining Journal*, Aug. 1, 8, 1885; Virginia City, *Evening Chronicle*, July 20, 30, Aug. 5, Sept. 3, 1885.

66. Hailey, *Wood River Times*, May 12, 1885.

67. San Francisco, *Mining and Scientific Press*, Nov. 26, 1887, March 24, 31, 1888; Nevada City, Calif., *Daily Transcript*, March 16, 18, 20, 25, 27, 30, April 16, 18, 20, 27, 1888.

THE GIBRALTAR OF UNIONISM

1. Butte, *Inter Mountain*, Holiday ed., Jan. 1, 1886; Harry C. Freeman, *A Brief History of Butte, Montana: The World's Greatest Mining Camp* (Chicago: Henry O. Shepard Co., 1900), pp. 35, 62; Jerre C. Murphy, *The Comical History of Montana: A Serious Story for a Free People* (San Diego: E. L. Scofield, 1912), p. 93.

2. Butte, *Daily Inter Mountain*, June 13, 1887; the history of the Butte Miners' Union is also treated by Vernon H. Jensen, *Heritage of Conflict: Labor Relations in the Non-Ferrous Metals Industry up to 1930* (Ithaca: Cornell University Press, 1950), pp. 289–354, and K. Ross Toole, "A History of the Anaconda Copper Mining Company: A Study in the Relationships between a State and Its People and a Corporation, 1880–1950," Unpublished Ph.D. Dissertation, University of California, Los Angeles, 1954, pp. 154 ff.

3. Butte, *Miner*, June 11, 18, July 2, 9, 1881; Butte, *Mining Journal*, Feb. 1, 1891; H. Knippenberg, *History of the Society of the Framers of the Constitution of the State of Montana* (Glendale, 1890), p. 81.

4. Butte, *Miner*, June 25, July 9, 30, 1878.

5. *Ibid.*, July 30, 1878.

6. *Ibid.*

7. *Ibid.*, Aug. 6, 1878.

8. *Ibid.*, June 21, 1881; Butte, *Daily Inter Mountain*, June 13, 1891.

9. Butte, *Daily Inter Mountain*, June 13, 1891.

10. *Ibid.*

11. Butte, *Miner*, Holiday ed., Jan. 1, 1889.

12. Butte, *Inter Mountain*, Holiday ed., Jan. 1, 1886; *Miner*, Holiday eds. Jan. 1, 1888, Jan. 1, 1889.

13. Butte, *Inter Mountain*, Holiday ed., Jan. 1, 1886.

14. *Ibid.*, Jan. 1, 1889, April 11, June 13, 1891.

15. *Ibid.*, June 13, 1887.

16. *Ibid.*, June 13, 14, 1887; Butte, *Mining Journal*, quoted in New York, *Engineering and Mining Journal*, Sept. 3, 1887.

17. Butte, *Daily Inter Mountain*, June 13, 1887.

18. New York, *Engineering and Mining Journal*, Aug. 13, Sept. 3, 1887.

19. *Ibid.*

20. Butte, *Daily Inter Mountain*, June 10, 12, 1891.

21. Butte, *Mining Journal*, Sept. 7, 10, 1890.

22. *Ibid.*, Nov. 23, 1890; New York, *Engineering and Mining Journal*, Dec. 6, 1890; *Proceedings and Debates of the Constitutional Convention Held in the City of Helena, Montana, July 4th 1889–August 17th 1889* (Helena: State Publishing Co., 1921), pp. 30, 213–214, 245.

23. Butte, *Mining Journal*, Feb. 22, 1891; *House Journal of the Second Session of the Legislative Assembly of the State of Montana, 5 January 1891 to 5 March 1891* (Helena: Journal Publishing Co., 1891), p. 6.

24. *Ibid.*, pp. 132, 155, 187.

25. Butte, *Mining Journal*, Jan. 21, Feb. 1, 8, 22, March 11, 1891.

26. *Ibid.*, Feb. 25, March 1, 4, 11, 1891.

27. *Ibid.*, March 25, May 11, June 10, 1891.

28. Butte, *Daily Inter Mountain*, June 10, 11, 18, 1891.

29. *Ibid.*, June 10, 12, 1891.

30. *Ibid.*, June 25, Aug. 10–Oct. 5, 1891; Butte, *Miner*, Jan. 1, 1889.

31. Butte, *Daily Inter Mountain*, Jan. 15, 1890, April 11, 1891; Denver, *Mining Industry and Tradesman*, Jan. 21, 1892; Butte, *Bystander*, July 1, 1893; Denver, *Miners' Magazine*, Feb. 1900, pp. 28–29.

32. Robert W. Smith, *The Coeur d'Alene Mining War of 1892: A Case Study of an Industrial Dispute* (Corvallis: Oregon State University Press, 1961); New York, *Engineering and Mining Journal*, April 11, 18, 25, 1891; Park City, Utah, *Record*, April 11, 18, 1891; Denver, *Mining Industry and Tradesman*, Aug. 5, 1887, June 3, 1891; *First Biennial Report of the Bureau of Labor Statistics of the State of Colorado, 1887–1888* (Denver, 1888), pp. 83, 102; *Third Biennial Report of the Bureau of Labor Statistics of the State of Colorado, 1891–1892* (Denver, 1892), pp. 59, 61; Charles M. Hough, "Leadville, Colorado, 1878 to 1898: A Study in Unionism," Unpublished M.A. Thesis, University of Colorado, 1958, pp. 61–77.

THE START OF THE WAR

1. Richard P. Rothwell, ed., *The Mineral Industry . . . 1893* (New York: Scientific Publishing Co., 1894), II, 311–312.

2. New York, *Engineering and Mining Journal*, Dec. 12, 1891; Candelaria, Nev., *Chloride Belt*, Nov. 21, Dec. 26, 1891.

3. Candelaria, Nev., *Chloride Belt*, Nov. 28, Dec. 5, 1891, March 12, 1892.

4. *Ibid.*, April 2, 9, 16, Dec. 24, 1892.

5. An excellent history of the Coeur d'Alene dispute is given in Robert Wayne Smith, *The Coeur d'Alene Mining War of 1892: A Case Study of an Industrial Dispute* (Corvallis: Oregon State University Press, 1961). Accounts by some of the participants are found in George Edgar French, "The Coeur d'Alene Riots of 1892," *Overland Monthly*, 26 (July 1895), 32–48; John Hays Hammond, *The Autobiography of John Hays Hammond* (New York: Farrar and Reinhart, 1935), I, 188–195; James H. Hawley, ed., *History of Idaho: The Gem of the Mountains* (Chicago: S. J. Clarke Publishing Co., 1920), I, 245–251; May A. Hutton, *The Coeur d'Alenes, or a Tale of the Modern Inquisition in Idaho* (Wallace, Idaho, 1900); Charles A. Siringo, *A Cowboy Detective* (Chicago, 1912), pp. 135 ff; Charles A. Siringo, *Two Evil Isms: Pinkertonism and Anarchism* (Chicago: Siringo, 1915), pp. 36 ff; Charles A. Siringo, *Riata and Spurs* (Boston: Houghton Mifflin Co., 1927), pp. 158–183; William T. Stoll, *Silver Strike* (Boston: Little,

Brown and Co., 1932). Other historical accounts have been written by Job Harriman, *The Class Struggle in Idaho* (New York: Labor Publishing Association, 1904) and serially in *Miners' Magazine*, Oct.–Nov. 1903; John M. Henderson, William S. Shiach, and Harry B. Averill, *An Illustrated History of North Idaho* (n.p.: Western Historical Publishing Co., 1903), pp. 1001–1008; and Vernon H. Jensen, *Heritage of Conflict* (Ithaca: Cornell University Press, 1950), pp. 25–37.

6. Smith, *op. cit.*, pp. 17–26.

7. *Ibid.*, pp. 28–30.

8. Spokane, Wash., *Spokane Review*, May 28, 1892; Siringo, *A Cowboy Detective*, p. 137, Two Evil Isms, p. 36, *Riata and Spurs*, p. 158; Hammond, *op. cit.*, p. 189; Henderson, *op. cit.*, p. 1003.

9. Smith, *op. cit.*, p. 31–33.

10. *Spokane Review*, March 19, 1892.

11. *Ibid.*, March 22, July 26, 1892; Smith, *op. cit.*, pp. 34–35.

12. *Spokane Review*, March 27, 1892, quoted in Henderson, *op. cit.*, pp. 1001–1004.

13. *Spokane Review*, March 30, 31, 1892.

14. *Ibid.*, April 6, 1892; Smith, *op. cit.*, p. 40.

15. *Spokane Review*, March 30, 1892; Smith, *op. cit.*, pp. 41–42; French, *op. cit.*, p. 33.

16. Smith, *op. cit.*, pp. 42–43.

17. *Spokane Review*, May 15, 17, 1892; Smith, *op. cit.*, pp. 44–46.

18. *Spokane Review*, May 15, 1892; Smith, *op. cit.*, pp. 44–45; Siringo, *Riata and Spurs*, p. 160.

19. *Spokane Review*, May 17, 18, 1892; *Miner's Magazine*, Oct. 22, 1903; Smith, *op. cit.*, p. 45.

20. *Spokane Review*, May 18, 1892; Smith, *op. cit.*, p. 46.

21. *Ibid.*, pp. 39, 53–54.

22. *Ibid.*, pp. 47–49; *Spokane Review*, May 22, June 5, 1892.

23. *Ibid.*

24. *Ibid.*, July 12, 1892.

25. *Ibid.*; Siringo, *Riata and Spurs*, pp. 164–172; Smith, *op. cit.*, pp. 62–63.

26. Spokane, Wash., *Spokane Daily Chronicle*, July 11, 1892; *Spokane Review*, July 12, 15, 1892; Smith, *op. cit.*, pp. 65–67.

27. *Spokane Review*, July 14, 1892; Smith, *op. cit.*, p. 65.

28. *Spokane Review*, July 13, 1892; Smith, *op. cit.*, p. 65.

29. *Spokane Review*, July 12, 13, 1892.

30. *Ibid.*, July 12, 1892.

31. *Ibid.*; *French*, op. cit., pp. 37–38; Smith, *op. cit.*, pp. 74–78.

32. French, *op. cit.*, pp. 39–40; Smith, *op. cit.*, p. 79.

33. *Spokane Review*, July 16, 1892, Smith, *op. cit.*, pp. 80–81.

34. *Ibid.*; *Spokane Daily Chronicle*, July 15, 16, 1892.

35. *Spokane Review*, July 16, 17, 19, 22, 1892; Smith, *op. cit.*, pp. 85–85.

36. *Spokane Review*, July 21, 24, 1892; French, *op. cit.*, pp. 43–45; Smith, *op. cit.*, pp. 86–89.

37. *Spokane Review*, July 22, 23, 1892; French, *op. cit.*, p. 43; Smith, *op. cit.*, pp. 87–88.

38. *Spokane Review*, July 21, 26, 1892; Smith, *op. cit.*, p. 88.

39. *Spokane Review*, July 22, 1892; *Spokane Daily Chronicle*, July 29, 1892.

40. Smith, *op. cit.*, pp. 83–84.

41. *Ibid.*, pp. 98–99.

42. *Ibid.*, pp. 99–100.

43. *Ibid.*, pp. 102–103, 122; San Francisco, *Mining and Scientific Press*, Sept. 17, 1892.

44. Smith, *op. cit.*, pp. 103–105.

45. *Spokane Review*, July 19, 21, 22, 1892; French, *op. cit.*, p. 44.

46. *Spokane Review*, July 24, 1892.

47. Butte, *Bystander*, Jan. 14, 1893; Smith, *op. cit.*, p. 84.

48. Butte, *Bystander*, Jan. 14, March 11, April 1, 1893.

49. *Ibid.*, March 11, 1893.

50. *Ibid.*, April 15, 1893; Smith, *op. cit.*, p. 116.

51. Butte, *Bystander*, Aug. 12, 26, Oct. 14, 1893; Smith, *op. cit.*, p. 119.

52. Butte, *Bystander*, May 20, 1893; New York, *Engineering and Mining Journal*, Feb. 4, 11, 25, 1893; San Francisco, *Mining and Scientific Press*, Feb. 18, 25, 1893; Denver, *Mining Industry*, Jan. 12, 26, 1893.

53. *Ibid.*, Jan. 19, 26, 1893.

54. *Ibid.*, March 30, 1893; Park City, Utah, *Record*, March 4, 11, 25, April 8, 1893; Butte, *Bystander*, April 29, June 10, 1893; Beth Kay Harris, *The Towns of Tintic* (Denver: Sage Books, 1961), pp. 139–142.

55. Denver, *Mining Industry*, Feb. 16, 1893.

56. Denver, *Miners' Magazine*, June, 1901.

THE FEDERATION

1. Anaconda, Mont., *Anaconda Standard*, May 16, 1893; Selig Perlman and Philip Taft, *History of Labor in the United States*. Vol. IV, *Labor Movements 1896–1932* (New York: Macmillan Co., 1935), pp. 172–173.

2. New York, *Engineering and Mining Journal*, Aug. 8, 1903, Sept. 15, 1904; "Western Federation of Miners," *Outlook*, 83 (July 7, 1906), 554–555 (italics added).

3. William Hard, "The Western Federation of Miners," *Outlook*, 83 (May 19, 1906), 133.

4. Anaconda, Mont., *Anaconda Standard*, May 13, 16, 1893; Butte, *Bystander*, May 13, 1893.

5. *Ibid.*; "Proceedings of the First Annual Convention of the Western Federation of Miners, Held at Butte, Montana, May 15–19, 1893," Library of the University of California, Berkeley.

6. *Ibid.*, pp. 3–15.

7. *Ibid.*, p. 6; *Constitution, By-Laws, Order of Business and Rules of Order of the Miners' Union of Virginia, Nevada* (Virginia: Brown and Mahanny, 1879), p. 5; *Constitution, By-Laws, Order of Business and Rules of Order of the Miners' Union of Gold Hill, Nev.* (Virginia City: Enterprise, 1871), p. 3.

8. "Proceedings . . . ," pp. 4–5.

9. Anaconda, Mont. *Anaconda Standard*, May 19, 1893; Butte, *Bystander*, May 20, 1893.

10. "Proceedings . . . ," pp. 8, 16–17.

11. *Ibid.*, p. 9.

12. *Ibid.*

13. *Ibid.*, pp. 18–20; Butte, *Bystander*, May 20, 1893; Anaconda, Mont., *Anaconda Standard*, May 21, 1893.

14. "Minutes of the Executive Committee of the Western Federation of Miner, 1893–1894" pp. 1–4, Library of the University of California, Berkeley.

15. Richard P. Rothwell, ed., *The Mineral Industry . . . 1893* (New York: Scientific Publishing Co., 1894), II, 311–312.

16. New York, *Engineering and Mining Journal*, Sept. 9, 23, Oct. 7, 14, Nov. 11, 1893; San Francisco, *Mining and Scientific Press*, Sept. 9, 23, Oct. 21, 1893.

Bibliography

Bibliography

MANUSCRIPTS

Cross, Ira B. California Labor Notes 1847–1885; 1886–1890; 1891–1896; Unions and Protective Associations. Bancroft Library, University of California, Berkeley.

Gold Hill Miners' Union. Organizational Papers 1866–1875. (Constitution, by-laws, pledge and minutes of meetings December 13, 1866–August 13, 1868; November 18, 1872–May 10, 1875.) Special Collections Library, University of Nevada, Reno.

———. Organizational Papers 1883–1915. (Membership registers, minutes of meetings and account books, 15 vols.) Western Americana Collection, Beinecke Rare Book and Manuscript Library, Yale University, New Haven.

———. Organizational Papers 1886–1892. (Alphabetical list of members, monthly dues ledger and check stubs, 4 vols.) Bancroft Library, University of California, Berkeley.

Gould & Curry Silver Mining Company. Records 1859–1885. California Section, California State Library, Sacramento.

Mechanics' Union of Storey County. Agreement entered into between the mining superintendents and the mechanics' union, both of Storey County, State of Nevada, July 17th, A.D., 1878, detailing mechanical labor, its hours and wages. Special Collections Library, University of Nevada, Reno.

Western Federation of Miners. Proceedings of the First, Second and Third Annual Conventions, and Executive Board Minutes of the Western Federation of Miners, 1893–1895. University of California Library, Berkeley.

NEWSPAPERS AND PERIODICALS

Aspen, Colo., *Union Era*, 1891–1892.
Aurora, Nev., *Esmeralda Herald*, 1879, 1882.
Austin, Nev., *Reese River Reveille*, 1864, 1868, 1872, 1875–1881.
Belmont, Nev., *Courier*, 1876.
Bodie, Calif., *Daily Free Press*, 1880–1881.
Bodie, Calif., *Morning News*, 1879.
Bodie, Calif., *Weekly Standard*, 1878–1879.
Bodie, Calif., *Weekly Standard-News*, 1881.
Boise, Idaho, *Tri-Weekly Statesman*, 1885.
Butte, Mont., *Bystander*, 1893.
Butte, Mont., *Daily Inter Mountain*, 1887–1891.

Butte, Mont., *Miner*, 1878, 1881, Holiday editions 1888, 1889.
Butte, Mont., *Mining Journal*, 1890–1891.
Calico, Calif., *Print*, 1885.
Candelaria, Nev., *Chloride Belt*, 1891–1892.
Candelaria, Nev., *True Fissure*, 1881–1885.
Central City, Colo., *Daily Register*, 1873.
Deadwood, S.D., *Black Hills Daily Times*, 1877–1878.
Deadwood, S.D., *Black Hills Weekly Pioneer*, 1877–1882.
Denver, Colo., *Miners' Magazine*, 1900–1904.
Denver, Colo., *Mining Industry & Tradesman*, 1887–1893.
Denver, Colo., *Rocky Mountain News*, 1873.
Downieville, Calif., *Mountain Messenger*, 1881–1883.
Eureka, Nev., *Daily Leader*, 1881, 1884–1885.
Eureka, Nev., *Daily Sentinel*, 1876–1881, 1884–1887, 1889–1893.
Gold Hill, Nev., *Daily News*, 1864, 1866–1869, 1872, 1874, 1877–1878.
Grass Valley, Calif., *Foothill Tidings*, 1877.
Grass Valley, Calif., *Daily National*, 1868–1870.
Grass Valley, Calif., *Republican*, 1871–1872.
Hailey, Idaho, *Wood River Times*, 1881–1889.
Hamilton, Nev., *Inland Empire*, 1869–1870.
Independence, Calif., *Inyo Independent*, 1878, 1881.
Jackson, Calif., *Amador Dispatch*, 1870–1871.
Jackson, Calif., *Amador Ledger*, 1870–1871.
Leadville, Colo., *Daily Democrat*, 1880.
Leadville, Colo., *Weekly Herald*, 1880.
Los Angeles, Calif., *Daily Herald*, May 25, 1881.
Los Angeles, Calif., *Mining Review*, March 16, 1907.
Lundy, Calif., *Homer Mining Index*, 1880–1881.
Nevada City, Calif., *Gazette*, 1869.
Nevada City, Calif., *Daily Transcript*, 1877–1879, 1881, 1888.
New York, *Engineering and Mining Journal*, 1878–1902.
New York, *Mining Record*, 1880–1881.
New York, *Times*, 1877, 1880–1882, 1892.
New York, *Tribune*, 1880, 1892.
North San Juan, Calif., *Times*, 1874–1880.
Park City, Utah, *Record*, 1880–1881, 1891, 1893.
Pioche, Nev., *Record*, 1872–1873, 1876–1877.
Ruby Hill, Nev., *Mining News*, 1883–1884.
Sacramento, Calif., *Reporter*, 1871.
Sacramento, Calif., *Daily Union*, 1864, 1871.
Salt Lake City, Utah, *Daily Tribune*, 1881, 1885.
San Bernardino, Calif., *Times*, 1876.
San Diego, Calif., *Union and Bee*, March 1, 1889.
San Francisco, *Alta California*, 1869, 1871, 1878, 1882, 1884.
San Francisco, *Bulletin*, 1869, 1871.
San Francisco, *Chronicle*, 1871.
San Francisco, *Mining and Scientific Press*, 1862–1906.
San Jose, Calif., *Daily Herald*, 1868.
Shermantown, Nev., *White Pine Telegram*, 1869.
Silver City, Idaho, *Owyhee (Idaho) Avalanche*, 1865–1886.

Silver Reef, Utah, *Miner,* 1879–1883.
Spokane, Wash., *Daily Chronicle,* 1892.
Spokane, Wash., *Review,* 1892.
Treasure City, Nev., *White Pine News,* 1869–1870.
Tucson, Ariz., *Arizona Weekly Citizen,* 1884.
Tucson, Ariz., *Daily Star,* 1884.
Tuscarora, Nev., *Times-Review,* 1880.
Unionville, Nev., *Humboldt Register,* 1864–1869.
Virginia City, Nev., *Chronicle,* 1877, 1885, July 5, 1917.
Virginia City, Nev., *Territorial Enterprise,* 1869–1870, 1872, 1874, 1877–1879, 1881–1882.
Virginia City, Nev., *Daily Trespass,* 1867.
Virginia City, Nev., *Daily Union,* 1864.

GOVERNMENT PUBLICATIONS

California. Bureau of Labor Statistics. *Biennial Reports 1883–1894.* Sacramento: State Printer, 1884–1894.
———. Office of the Adjutant General. *Report . . . for the Years 1870 and 1871.* Sacramento: State Printer, 1871.
Colorado. Bureau of Labor Statistics. *Biennial Reports 1887–1894.* Denver, 1888–1894.
———. Office of the Adjutant General. *Biennial Report . . . for the Term Ending December 31, 1880.* Denver: Tribune Publishing Co., 1881.
Montana. Constitutional Convention. *Proceedings and Debates of the Constitutional Convention Held in the City of Helena, Montana, July 4th 1889–August 17th 1889.* Helena: State Publishing Co., 1921.
———. Legislative Assembly. *House Journal of the Second Session . . . 5 January 1891 to 5 March 1891.* Helena: Journal Publishing Co., 1891.
Nevada. State Legislature. *Journals of the Senate and Assembly, Fifth Session.* Carson City, 1870.
———. State Mineralogist. *Biennial Report . . . for the Years 1871 and 1872.* Carson City, 1873.
U.S. Bureau of the Census. *Eighth Census, 1860, I Population; II Manufactures.* Washington: G.P.O., 1864; 1865.
———. *Ninth Census, 1870, I Population; III Wealth and Industry.* Washington: G.P.O., 1872.
———. *Tenth Census, 1880, I Population; XIII Precious Metals; XIV Mining Laws.* Washington: G.P.O., 1883; 1885.
———. *Eleventh Census, 1890, I Population; VII Mineral Industries.* Washington: G.P.O., 1892; 1895.
U.S. Bureau of the Mint. *Report . . . upon the Statistics of the Production of the Precious Metals in the United States, 1880; 1881; 1882; 1883; 1884.* Washington: G.P.O., 1881–1885.
U.S. Commission on Industrial Relations. *Final Report and Testimony Submitted to Congress by the Commission on Industrial Relations,* Vol. IV. Washington: G.P.O., 1916.
U.S. Commissioner of Labor. *Special Report, Labor Laws of the United States.* Washington: G.P.O., 1896.

————. *Third Annual Report, 1887, Strikes and Lockouts.* Washington: G.P.O., 1888.

U.S. Congress. House. *Coeur d'Alene Mining Troubles* (56th Cong., 1st Sess., H. Rept. 1999). Washington: G.P.O., 1900. 132 pp.

————. Senate. *Coeur d'Alene Mining Troubles* (56th Cong., 1st Sess., S. Docs. 24; 25; 42. Washington: G.P.O., 1899.

————. ————. *A Report on Labor Disturbances in the State of Colorado, from 1880 to 1904 Inclusive* (58th Cong., 3d Sess., S. Rept. 122). Washington: G.P.O., 1905, 363 pp.

U.S. Industrial Commission. *Report, 1901, XII Capital and Labor Employed in the Mining Industry.* Washington: G.P.O., 1901. 747 pp.

U.S. Secretary of the Interior. *Report . . . 1884* (48th Cong., 2d Sess. Ex. Doc. 1, Pt. 5). Washington: G.P.O., 1884.

LABOR ORGANIZATION PUBLICATIONS

Globe City Union. *Constitution and By-Laws of the Globe City Union of Globe, Arizona, Organized May 14th, 1884.* Globe: Arizona Silver Belt Print, 1884. 24 pp.

Gold Hill Miners' Union. *Constitution, By-Laws, Order of Business and Rules of Order of the Miners' Union of Gold Hill, Nev.* Virginia City: Enterprise, 1871. 10 pp.

————. *Constitution, By-Laws, Order of Business and Rules of Order of the Miners' Union of Gold Hill, Nev.* Virginia, Nev.: D. L. Brown, 1885. 21 pp.

Grass Valley Miners' Union. *Constitution, By-Laws, Order of Business and Rules of Order of the Miners' Union of Grass Valley, Cal., Organized May 1869.* Grass Valley: Daily National, 1869. 20 pp.

Virginia City Miners' Union. *Constitution, By-Laws, Order of Business and Rules of Order of the Miners' Union of Virginia, Nevada.* Virginia: Brown and Mahanny, 1879. 20 pp.

Virginia City Miners' Union. *Constitution and By-Laws of the Virginia City, Nevada, Miners' Union No. 46 of the Western Federation of Miners.* Reno: Nevada Press, 1911. 24 pp.

Washoe Typographical Union. *Constitution and By-Laws of the Washoe Typographical Union, Including the Scale of Prices and List of Members. Organized June 28, 1863.* Virginia, Nevada Terr.: Standard Book and Job Printing Office, 1863.

Western Federation of Miners. *Constitution and By-Laws of the Western Federation of Miners Adopted at Butte City, Montana, May 19, 1893. Amended at Salt Lake City, Utah, 1894. Amended at Denver, Colorado, 1895.* Butte, Mont.: Bystander printers, 1895. 23 pp.

THESES AND DISSERTATIONS

Be Dunnah, Gary P. "A History of the Chinese in Nevada: 1855–1904." Unpublished M.A. Thesis, University of Nevada, Reno, 1966.

Cash, Joseph Harper. "Labor in the West: The Homestake Mining Company and Its Workers, 1877–1942." Unpublished Ph.D. Dissertation. University of Iowa, 1966. 277 pp.

Hough, Charles Merrill. "Leadville, Colorado, 1878 to 1898: A Study in Unionism." Unpublished M.A. Thesis, University of Colorado, 1958. 110 pp.

Lee, Rose Hum. "The Growth and Decline of Chinese Communities in the Rocky Mountain Region." Unpublished Ph.D. Dissertation. University of Colorado, 1947. 371 pp.

Love, Alice Emily. "The History of Tombstone to 1887." Unpublished M.A. Thesis, University of Arizona, 1933.

Rudolph, Gerald E. "The Chinese in Colorado, 1869–1911." Unpublished M.A. Thesis, University of Denver, 1964. 182 pp.

Smith, Duane Allan. "Silver Camp Called Caribou." Unpublished M.A. Thesis, University of Colorado, 1961.

Toole, Kenneth Ross. "A History of the Anaconda Copper Mining Company: A Study in the Relationship between a State and Its People and a Corporation, 1880–1950." Unpublished Ph.D. Dissertation, University of California, Los Angeles, 1954.

Trull, Fern Coble. "The History of the Chinese in Idaho from 1864 to 1910." Unpublished M.A. Thesis, University of Oregon, 1946.

Wyman, Walker de Marquis, Jr. "The Underground Miner, 1860–1910: Labor and Industrial Change in the Northern Rockies." Unpublished Ph.D. Dissertation, University of Washington, 1971. 403 pp.

ARTICLES

Bechtol, Paul T., Jr. "The 1880 Labor Dispute in Leadville," *Colorado Magazine*, 47 (1970), 312–325.

Bonner, J. "Leadville," *Lippincott's Magazine*, 24 (Nov. 1879), 604–615.

Burrows, A. "Social Life among Western Miners," *Sunset*, 16 (March 1906), 434–435.

Church, John A. "Accidents in the Comstock Mines and Their Relation to Deep Mining," *Transactions of the American Institute of Mining Engineers*, 8 (1879), 84–97.

Crane, Paul, and Alfred Larson. "The Chinese Massacre," *Annals of Wyoming*, 12 (1940), 47–55.

Dubofsky, Melvyn. "The Leadville Strike of 1896–97: An Appraisal," *Mid-America*, 48 (1966), 99–118.

————. "The Origins of Western Working Class Radicalism, 1890–1905," *Labor History*, 7 (1966), 131–154.

Earl, Phillip I. "Nevada's Italian War, July–September 1879," *Nevada Historical Society Quarterly*, 12 (1969), 51–87.

Fahey, John. "Coeur d'Alene Confederacy," *Idaho Yesterdays*, 12 (1968), 2–7.

Foote, M. "A California Mining Camp," *Scribners Monthly*, 15 (1878), 480–493.

Freeland, Francis T. "Mining Leases," *Transactions of the American Institute of Mining Engineers*, 25 (1895), 106–112.

French, George Edgar. "The Coeur d'Alene Riots of 1892," *Overland Monthly*, 26 (July 1895), 32–48.

Gaboury, William J. "From Statehouse to Bull Ben: Idaho Populism and the Coeur d'Alene Troubles of the 1890's," *Pacific North West Quarterly*, 58 (1967), 4–22.

Hand, Wayland. "The Folklore, Customs, and Traditions of the Butte Miner," *California Folklore Quarterly*, 5 (1946), 1–25, 153–176.

Hanson, Harry M. "Gold Mining in the Black Hills," *Engineering Magazine*, 3 (1892), 687–695.

Hard, William. "The Western Federation of Miners," *Outlook*, 83 (May 19, 1906), 125–133.

Harriman, Job. "The Class Struggle in Idaho," *Miners' Magazine* (Denver), October 8, 15, 22, 1903.

Hough, Merrill. "Leadville and the Western Federation of Miners," *Colorado Magazine*, 49 (1972), 19–34.

Lawrence, Benjamin B. "Notes on the Lease or Tribute System of Mining, as Practiced in Colorado," *Transactions of the American Institute of Mining Engineers*, 21 (1892–1893), 911–919.

Levinne, Marvin J. "The Homestake Mining Company: The Strategy and Semantics of Organizational Conflict 1877–1963," *South Dakota, Department of History, Reports and Historical Collection*, 33 (1966), 512–541.

Morefield, Richard H. "Mexicans in the California Mines, 1848–1853," *California Historical Society Quarterly*, 35 (1956), 37–46.

Ourada, Patricia K. "The Chinese in Colorado," *Colorado Magazine*, 29 (1952), 273–284.

Owens, Kenneth. "Pierce City Incident 1885–1886," *Idaho Yesterdays*, 3 (Fall 1959), 8–13.

Pawar, S. B. "The Structure and Nature of Labor Unions in Utah, 1890–1920," *Utah Historical Quarterly*, 35 (1967), 236–255.

Pendleton, Mark A. "Memories of Silver Reef," *Utah Historical Quarterly*, 3 (Oct. 1930), 99–118.

Perrigo, Lynn I. "Cornish Miners of Early Gilpin County," *Colorado Magazine*, 14 (1937), 92–101.

Raymond, Rossiter W. "The Hygiene of Miners," *Transactions of the American Institute of Mining Engineers*, 8 (1879), 108–109.

Sander, Helen F. "Butte—The Heart of the Copper Industry," *Overland Monthly*, n.s. 48 (1906), 367–384.

Saunders, W. L. "History of Rock Drills," *Mining and Scientific Press* (San Francisco), May 21, 1910.

Smith, Duane A. "The Caribou—a Forgotten Mine," *Colorado Magazine*, 39 (1962), 47–54.

———. "Colorado's Urban-Mining Safety Valve," *Colorado Magazine*, 48 (1971), 299–318.

Suggs, George G., Jr. "Catalyst for Industrial Change: The WFM, 1893–1903," *Colorado Magazine*, 45 (1968), 322–339.

Thomason, Frank. "The Bellevue Stranglers," *Idaho Yesterdays*, 13 (Fall 1969), 26–32.

Wells, Merle W. "The Western Federation of Miners," *Journal of the West*, 12 (1973), 18–35.

Williams, Albert. "Modern Types of Gold and Silver Miners," *Engineering Magazine*, 2 (Oct. 1891), 48–62.

BOOKS

Adamic, Louis. *Dynamite: The Story of Class Violence in America*. New York: Viking Press, 1934. 495 pp.

Amador Mining Company. *Annual Report . . . 1870*. San Francisco: Stock Report, 1871.

———. *Annual Report . . . 1871*. San Francisco: Stock Report, 1872.

Angel, Myron, ed. *History of Nevada*. Oakland, Calif.: Thompson and West, 1881. 680 pp.

Ayers, James J. *Gold and Sunshine: Reminiscences of Early California*. Boston: Richard G. Badger, 1922. 359 pp.

Bancroft, Caroline. *Gulch of Gold: A History of Central City, Colorado*. Denver: Sage Books, 1958. 387 pp.

Bancroft, Hubert H. *History of Arizona and New Mexico, 1530–1888*. San Francisco: The History Co., 1890. 829 pp.

———. *History of California*. San Francisco: The History Co., 1884–1890. 7 vols.

———. *History of Nevada, Colorado and Wyoming, 1540–1888*. San Francisco: The History Co., 1890. 828 pp.

———. *History of Washington, Idaho and Montana, 1845–1889*. San Francisco: The History Co., 1890. 836 pp.

Barth, Gunther. *Bitter Strength: A History of the Chinese in the United States, 1850–1870*. Cambridge: Harvard University Press, 1964. 305 pp.

Barton, D. B. *Essays in Cornish Mining History*. Truro: D. Bradford Barton, 1968. 198 pp.

Beal, Merrill D., and Merle W. Wells. *History of Idaho*. New York: Lewis Historical Publishing Co., 1959. 3 vols.

Bennett, Estelline. *Old Deadwood Days*. New York: J. H. Sears and Co., 1928. 300 pp.

Blake, William P. *Notices of Mining Machinery and Various Mechanical Appliances in Use Chiefly in the Pacific States and Territories for Mining, Raising and Working Ores*. New Haven: Charles Chatfield and Co., 1871. 245 pp.

———. *Tombstone and Its Mines*. New York: Cheltenham Press, 1902. 82 pp.

Brissenden, Paul F. *The I. W. W.: A Study of American Syndicalism*. New York: Russell and Russell, 1957. 438 pp.

Bromley, Isaac C. *The Chinese Massacre at Rock Springs, Wyoming Territory, September 2, 1885*. Boston: Franklin Press, 1886. 92 pp.

Brown, George R., ed. *Reminiscences of Senator William M. Stewart of Nevada*. New York and Washington: Neale Publishing Co., 1908. 358 pp.

Buchanan, Joseph R. *The Story of a Labor Agitator*. New York: Outlook Co., 1903. 460 pp.

Burlingame, Merrill G., and K. Ross Toole. *History of Montana*. New York: Lewis Historical Publishing Co., 1957. 3 vols.

Chiu, Ping. *Chinese Labor in California, 1850–1880: An Economic Study.* Madison: State Historical Society of Wisconsin, 1963. 180 pp.

Corbett, Thomas B., and John H. Ballanger. *Second Annual Leadville City Directory.* Leadville, 1881.

Cross, Ira B. *A History of the Labor Movement in California.* Berkeley: University of California Press, 1935. 354 pp.

Curtis, Joseph S. *Silver-Lead Deposits of Eureka, Nevada.* Washington: G.P.O., 1884. 200 pp.

Davis, Carlyle C. *Olden Times in Colorado.* Los Angeles: Phillips Publishing Co., 1916. 448 pp.

Dill, R. G. *The Political Campaigns of Colorado.* Denver: Arapahoe Publishing Co., 1895. 286 pp.

Dubofsky, Melvyn. *We Shall All Be: A History of the Industrial Workers of the World.* Chicago: Quadrangle Books, 1969. 557 pp.

Finlay, James Ralph. *The Cost of Mining: An Exhibit of the Results of Important Mines Throughout the World.* New York: McGraw-Hill Book Co., 1910. 415 pp.

Freeman, Harry C. *A Brief History of Butte, Montana: The World's Greatest Mining Camp.* Chicago: Henry O. Shepard Co., 1900. 123 pp.

Griswold, Don L., and Jean H. Griswold. *The Carbonate Camp Called Leadville.* Denver: University of Colorado Press, 1951. 282 pp.

Hammond, John Hays. *The Autobiography of John Hays Hammond.* New York: Farrar and Reinhart, 1935. 2 vols.

Harpending, Asbury. *The Great Diamond Hoax and Other Stirring Incidents in the Life of Asbury Harpending.* San Francisco: James H. Barry Co., 1913. 283 pp.

Harriman, Job. *The Class Struggle in Idaho.* New York: Labor Publishing Assoc., 1904.

Harris, Beth Kay. *The Towns of Tintic.* Denver: Sage Books, 1961. 180 pp.

Hawley, James H., ed. *History of Idaho: The Gem of the Mountains.* Chicago: S. J. Clarke Publishing Co., 1920. 4 vols.

Henderson, Charles H. *Mining in Colorado: A History of Discovery, Development and Production.* Washington: G.P.O., 1926. 263 pp.

Henderson, John M., William S. Shiach, and Harry B. Averill. *An Illustrated History of North Idaho.* N.p.: Western Historical Publishing Co., 1903. 1,238 pp.

Hinton, Richard J. *The Handbook to Arizona.* San Francisco: Payot, Upham and Co., 1878. 431 + ci pp.

History of the Arkansas Valley, Colorado. Chicago: O. L. Baskin and Co., 1881. 889 pp.

Hubbard, Lester A. *Ballads and Songs from Utah.* Salt Lake City: University of Utah Press, 1961. 475 pp.

Hutton, May A. *The Coeur d'Alenes, or, a Tale of the Modern Inquisition in Idaho.* Wallace, Idaho: M. A. Hutton, 1900. 246 pp.

Jackson, W. Turrentine. *Treasure Hill: Portrait of a Silver Mining Camp.* Tucson: University of Arizona Press, 1963. 254 pp.

Jenkins, A. K. H. *The Cornish Miner, An Account of His Life Above and Underground From Early Times.* London: George Allen and Unwin Ltd., 1927.

Jensen, Vernon H. *Heritage of Conflict: Labor Relations in the Nonferrous Metals Industry up to 1930.* Ithaca: Cornell University Press, 1950. 495 pp.

Knippenberg, Henry. *History of the Society of the Framers of the Constitution of the State of Montana.* Glendale, 1890. 154 pp.

Koontz, John. *Political History of Nevada.* Carson City, 1965. 175 pp.

Lord, Eliot. *Comstock Mining and Miners.* Washington: G.P.O., 1883. 451 pp.

Malloy, William M., compiler. *Treaties, Conventions, International Acts, Protocols and Agreements between the United States of America and Other Powers, 1776–1909.* Washington: G.P.O., 1910.

Mason, Jesse D. *History of Amador County.* Oakland: Thompson and West, 1881. 344 pp.

A Memorial and Biographical History of Northern California. Chicago: Lewis Publishing Co., 1891. 834 pp.

Murphy, Jerre C. *The Comical History of Montana: A Serious Story for a Free People.* San Diego: E. L. Scofield, 1912. 332 pp.

Myers, John M. *The Last Chance: Tombstone's Early Years.* New York: Dutton and Co., 1950. 260 pp.

Myers, Margaret G. *A Financial History of the United States.* New York: Columbia University Press, 1970. 451 pp.

Paul, Rodman W. *California Gold: The Beginning of Mining in the Far West.* Cambridge: Harvard University Press, 1947. 380 pp.

———. *Mining Frontiers of the Far West 1848–1880.* New York: Holt, Rinehart and Winston, 1963. 236 pp.

Perlman, Selig, and Philip Taft. *History of Labor in the United States.* Vol. IV, *Labor Movements 1896–1932.* New York: Macmillan Co., 1935.

Pickett, Charles E. *A Letter from Charles E. Pickett to Jno. A. Eagan.* San Francisco: White and Bauer, 1871. 8 pp.

Raymond, Rossiter W. *Statistics of Mines and Mining . . . for the Year 1869 through 1871.* Washington: G.P.O., 1870–1872. 4 vols.

Reed, Jewett V., and A. K. Harcourt. *The Essentials of Occupational Diseases.* Springfield, Ill.: Charles C. Thomas, 1941. 225 pp.

Rickard, Thomas A. *A History of American Mining.* New York: McGraw-Hill, 1932. 419 pp.

Rothwell, Richard P. *The Mineral Industry . . . 1893.* New York: Scientific Publishing Co., 1894. 894 pp.

Sandmeyer, Elmer C. *The Anti-Chinese Movement in California.* Urbana: University of Illinois Press, 1939. 127 pp.

Saxton, Alexander P. *The Indispensable Enemy: Labor and the Anti-Chinese Movement in California.* Berkeley and Los Angeles: University of California Press, 1971. 293 pp.

Shinn, Charles H. *The Story of the Mines, as Illustrated by the Great Comstock Lode of Nevada.* New York: D. Appleton & Co., 1896. 272 pp.

Siringo, Charles A. *A Cowboy Detective.* Chicago: Conkey, 1912. 519 pp.

———. *Riata and Spurs.* Boston: Houghton Mifflin Co., 1927. 276 pp.

———. *Two Evil Isms: Pinkertonism and Anarchism.* Chicago: Siringo, 1915. 109 pp.

Smith, Duane A. *Horace Tabor: His Life and the Legend.* Boulder: Colorado Associated University Press, 1973.

Smith, Grant H. *The History of the Comstock Lode 1850–1920*. Reno, 1943 University of Nevada Bulletin, vol. 37, no. 3.

Smith, Robert Wayne. *The Coeur d'Alene Mining War of 1892: A Case Study of an Industrial Dispute*. Corvallis: Oregon State University Press, 1961. 132 pp.

Spence, Clark C. *British Investment and the American Mining Frontier 1860–1901*. Ithaca, N.Y.: Cornell University Press, 1958. 288 pp.

————. *Mining Engineers and the American West: The Lace-Boot Brigade, 1849–1933*. New Haven: Yale University Press, 1970. 407 pp.

Stoll, William T. *Silver Strike*. Boston: Little, Brown and Co., 1932. 273 pp.

Suggs, George G. *Colorado's War on Militant Unionism: James H. Peabody and the Western Federation of Miners*. Detroit: Wayne State University Press, 1972. 242 pp.

Todd, Arthur C. *The Cornish Miner in America*. Truro: D. Bradford Ltd., 1967, 279 pp.

Waldorf, John Taylor. *A Kid on the Comstock*. Berkeley: Friends of Bancroft Library, 1968. 92 pp.

Webb, Sidney, and Beatrice Webb. *The History of Trade Unionism*. New York: Longmans, Green and Co., 1920.

Wells, Harry L., *History of Nevada County, California*. Oakland: Thompson and West, 1880. 234 pp.

Wieck, Edward A. *The American Miners' Association: A Record of the Origins of Coal Miners' Unions in the United States*. New York: Russell Sage Foundation, 1940. 330 pp.

Woodard, Bruce A. *Diamonds in the Salt*. Boulder: Pruett Press, 1967. 200 pp.

Wright, Morris. *"Takes More than Guns," A Brief History of the International Union of Mine, Mill and Smelter Workers*. Denver: I.U.M.M.S.W., 1944. 48 pp.

Wright, William (Dan De Quille). *History of the Big Bonanza*. Hartford: American Publishing Co., 1876. 569 pp.

Index

Absentee ownership, 4, 27–28, 31, 106, 121

Accidents underground, 23 ff.; blasting, 25–26; due to carbon dioxide, 15, 24; cave-in, 25; due to fires, 26; in inclines, 24; rate of, 23; in shafts, 23–24; union benefits for, 27

Ada County Jail, 212, 215, 218

Adit, 12

Affiliation of Nevada miners unions, 128–129, 133

Agents provocateurs, 66; Coeur d'-Alenes, 197, 199, 206; White Pine, 74–75

Albion Mining Co., 140

Alice mine, 182

Alleghany Miners' Union, 131

Allen, W. I., 158, 160–161

Allison, C. Leon. *See* Charles Siringo

Alpine Miners' Association, 142

Alta Mining Co., 56–58, 137

Amador City (Calif.), 90

Amador County Laborers' Association: organization and purpose of, 91, 118; political activity of, 91–92, 102–103, 125; strike by, 92–101

Amador Mining Co., 90–101 *passim*

"Amador War," 90 ff., 118, 206, 227

Amusements in mining camps, 11, 52–54, 133

Ancient Order of Hibernians, 53, 133

Anti-Chinese agitation, 107 ff.; in California, 117 ff.; in Colorado, 119; in Idaho, 118–119; in Montana, 124; in Nevada, 109–117, 123; role of miners' unions in, 56, 112–127; in Wyoming, 127

Anti-Chinese Club, 123

Anti-Chinese legislation, 111, 125

Anti-Coolie Association, 118

Apex mine, 10

Arbitration, 96, 175–176, 189, 222

Arizona, 6, 7; miners' unions in, 133, 164–169; miners' wages in, 164, 169

Arizona Silver Mining Co., 109

Arrests of union men at: Austin, 77; Butte, 193; Central City, 104; Coeur d'Alene, 210; Leadville, 154; Ruby Hill, 178; Silver Reef, 161; Unionville, 110–111; White Pine, 72; Wood River, 172–173, 176

Arthur, Pres. Chester A., 126

Aspen (Colo.): Miners' Protective Association, 142; Miners' Union, 194, 216, 220, 225

Atkinson, Thomas A., 56, 115–116

Atlanta (Ida.), 169

Auburn (Calif.), 121

Aurora (Nev.), 137, 139; Miners' Protective Association, 39; Miners' Union, 131

Aurora Consolidated Mining Co., 69–70, 75

Austin (Nev.), 105; Miners' league in, 39, 67; Miners' unions in, 76–77, 79, 128–130, 139, 159–160, 179; strikes in, 76–77, 130; wages in, 76, 134–135

Ayers, James J., 70–72, 75

Baldwin, A. W., 110–111

Baldwin, M. A., 136

Bank of California, 96

"Bank ring." *See* "Mill ring"

Bannack Miners' Union, 220, 224

Barbee & Walker Mine, 160–162

Barker Miners' Union, 194, 220

Barris, Lavern, 70

Bateman, W. C., 33

Batterman, C. M., 166–168

Battle Mountain Miners' Union, 133, 159

Bean, Ivory, 213

Bear Butte Miners' Union, 132

Beatty, James H., 211–212, 215

Beaver (Utah), 161

Bell, V. G., 120

Bellevue (Ida.), 169, 171–173, 175; stranglers in, 173, 175–176

Belt Mountain Miners' Union, 220

Benefits, union: death, 27, 51–52;

Tybo Workingmen's Protective Union, 123, 141
Typographical unions: at Butte, 187, 192; at Leadville, 148; at Washoe, 45, 113

Uncle Sam mine, 33–34
Underground mining: conditions, 13 ff.; dust, 16–17; fires, 26, 59; gases, 15–16; hazards 23 ff.; heat, 13–15; water, 13–14, 24
Union Milling & Mining Co., 64
Union mine, 202, 204
Union Pacific Railroad, 200
Unions. See Miners' Union; Mechanics' Union; and names of individual union
Unionville (Nev.): Chinese expulsion from, 54, 109–111; Workingman's Protective Union, 109–114
University mine, 140
Utah, 10, 16; miners' unions in, 132, 158–162, 194, 216–217, 220, 224–225; miners' wages in, 134, 160, 169, 216

Vagrancy law, used to break strike, 154
Van Arman, H. M., 168
Ventilation in mines, 12–13, 15–18, 28, 86
Vienna (Ida.), 169
Vigilantes: in Bellevue, 173, 175–176; in Leadville, 151–154; in Pioche, 78; in Silver Reef, 161
Violence in western mines, origin of, 227–228
Virginia & Truckee Railroad, 113–117
Virginia City (Nev.), 32 ff., 110–111
Virginia City Miners' Protective Association, 32–33, 39, 74
Virginia City Miners' Union, 69, 78, 91, 101, 128, 131, 133, 139, 148, 182, 220–222; aids other unions, 166, 173, 198; and anti-Chinese activity, 112–117, 126; benefits of, 52; constitution adopted by, 49, 52, 129, 221; enforces closed shop, 60–61, 129–130; hall, 53; library, 53–54; membership in, 49, 58; militancy of, 49; minimum wage, demands of, 50–51; organization of, 49; political activity of, 54–56, 101, 111; and strikes, 50–51
Visiting committee, 51

Wages of carmen, 42, 50, 84, 199, 201
Wages of engineers, 59, 92, 141
Wages of millmen, 38, 184
Wages of miners, 8, 19, 27, 42, 50, 61, 68, 75, 84, 92, 103–104, 131, 134, 201, 216, 225
  Chinese, 107–108, 120, 122
  in Comstock mines, 32 ff., 135, 157
  for drift work, 42, 183
  in gold camps, 81, 90, 92, 103, 134
  in hydraulic mines, 120–122, 134
  Mexican, 6
  minimum, 27, 37–39, 44, 47–48, 50, 61, 63–64, 75, 77, 85, 92, 121–122, 132–136, 147–151, 157 ff., 183, 199, 215, 222, 226
  payment of: in "bullion checks," 30, 134–135; in gold, 30, 37, 44; in greenbacks, 30, 37; in scrip, 29–30; in stock, 30, 59, 177; in "store orders," 28–29; in trade dollars, 30, 135
  reduction of: and community opposition to, 35–37, 62, 70, 163, 165, 183; and community support of, 75, 163; mine owners' reasons for, 134, 158, 165, 196–197, 201, 224; organized efforts of owners for, 33–34, 163–164
  seizure of mines for, 135–137
  in silver camps, 134, 157 ff.
  sliding scale of, 62, 197, 217, 225
  for wet work, 42, 62, 183
Wages of muckers, 19, 50, 84, 199, 201
Wages of smelter workers, 141
Wages of surface workers, 42, 90
Walker Bros., 182
Wallace (Ida.), 198, 202–204, 209–210, 212
Ward, J. L., 165, 167
Wardner (Ida.): 198–210 passim; Miners' Union, 194, 198, 203, 226
Wardner Junction (Ida.), 198, 209
War Eagle Mountain, 79–80, 118
Warren, Joel, 203–204, 213
Washington, 220
Washoe County Miners' League. See Miners' League, Washoe County
Washoe Typographical Union, 45, 113
Watt, William, 88
Western Federation of Miners, 45, 132, 194–195, 217–218, 219 ff.; adopts constitution, 221–224; early struggles of, 225–226; militancy of,

219–220, 226–227; organization of, 218, 220–221

White, John G. (Jack): early career of, 43; at Grass Valley, 43, 89–90; organizes Gold Hill Miners' Union, 43–50; at White Pine, 68, 69

White Pine (Nev.): Miners' Benevolent Association, 68–69; Miners' Union, 68–75; rush to, 67–68

Whorehouses, 11

Willey, Gov. Norman, 205, 208, 211

Williams, John, 114

Williams, John L., 220

Winters, J. B., 41

Winze, 12

Witter, Aaron C., 183, 190

Wolf Tone Guards, 152

Wood, Fremont, 212

Woodburn, William, 37, 40, 42, 49, 173

Wood River (Ida.): Knights of Labor, 177; miners' unions in, 132, 169–176; mines in, 16, 169; wages in, 164, 169–170, 176

Woodworth, J. M., 54, 111

Working conditions. *See* Hardrock miners, working conditions

Working hours, 17, 92, 114, 157, 180–181; eight-hour shifts during, 59, 63, 69, 77, 146; legislation affecting, 190–193, 224; strikes because of, 69, 77–78, 114, 141, 146 ff.

Workingman's Advocate (labor paper), 112

Workingman's Protective Union, Unionville, 109–114

Workingmen's Club of Darwin, 141–142

Workingmen's Convention, Virginia City, 112–114

Workingmen's Party of California, 119, 125

Workingmen's Protective Union, Tybo, 123, 141

Workingmen's Union, Butte, of 1878, 132, 183–185; of 1890, 187, 192–193

Wyoming, miners' unions, 127

Yellow Jacket mine, 32, 34, 47, 116; fire in, 26, 59

Yosemite (steamer), 95